MANAGING GLOBAL GENETIC RESOURCES

Livestock

MANAGING GLOBAL GENETIC RESOURCES

Livestock

Committee on Managing Global Genetic Resources:
Agricultural Imperatives

Board on Agriculture
National Research Council

NATIONAL ACADEMY PRESS
Washington, D.C. 1993

NATIONAL ACADEMY PRESS • 2101 Constitution Avenue, NW • Washington, DC 20418

This material is based on work supported by the U.S. Department of Agriculture, Agricultural Research Service, under Agreement No. 59-32U4-6-75. Additional funding was provided by Calgene, Inc.; Educational Foundation of America; Kellogg Endowment Fund of the National Academy of Sciences and the Institute of Medicine; Monsanto Company; Pioneer Hi-Bred International, Inc.; Rockefeller Foundation; U.S. Agency for International Development; U.S. Forest Service; W. K. Kellogg Foundation; World Bank; and Basic Science Fund of the National Academy of Sciences, contributors to which include the Atlantic Richfield Foundation, AT&T Bell Laboratories, BP America, Inc., Dow Chemical Company, E.I. duPont de Nemours & Company, IBM Corporation, Merck & Co., Inc., Monsanto Company, and Shell Oil Company Foundation. Dissemination activities were supported in part by the W. K. Kellogg Foundation.

Library of Congress Cataloging-in-Publication Data

Livestock / Committee on Managing Global Genetic Resources: Agricultural Imperatives, Board on Agriculture, National Research Council.
p. cm. — (Managing global genetic resources)
Includes bibliographical references (p.) and index.
ISBN 0-309-04394-8
1. Livestock—Germplasm resources. I. National Research Council (U.S.). Committee on Managing Global Genetic Resources: Agricultural Imperatives. II. Series.
SF105.3.L58 1993
636.08'21—dc20 93-16715
CIP

Printed in the United States of America

Committee on Managing Global Genetic Resources: Agricultural Imperatives

PETER R. DAY, *Chair,* Rutgers University
ROBERT W. ALLARD, University of California, Davis
PAULO DE T. ALVIM, Comissão Executiva do Plano da Lavoura Cacaueira, Brasil*
JOHN H. BARTON, Stanford University
FREDERICK H. BUTTEL, University of Wisconsin
TE-TZU CHANG, International Rice Research Institute, The Philippines (Retired)
ROBERT E. EVENSON, Yale University
HENRY A. FITZHUGH, International Livestock Center for Africa, Ethiopia
MAJOR M. GOODMAN, North Carolina State University
JAAP J. HARDON, Center for Genetic Resources, The Netherlands
DONALD R. MARSHALL, University of Sydney, Australia
SETIJATI SASTRAPRADJA, National Center for Biotechnology, Indonesia
CHARLES SMITH, University of Guelph, Canada
JOHN A. SPENCE, University of the West Indies, Trinidad and Tobago

Genetic Resources Staff

MICHAEL S. STRAUSS, *Project Director*
JOHN A. PINO, *Project Director*†
BRENDA BALLACHEY, *Staff Officer*‡
BARBARA J. RICE, *Project Associate and Editor*

*Executive Commission of the Program for Strengthening Cacao Production, Brazil.
†Through June 1990.
‡Through November 1989.

Subcommittee on Animal Genetic Resources

HENRY A. FITZHUGH, *Chair,* International Livestock Center for Africa, Ethiopia
ELIZABETH L. HENSON, Cotswold Farm Park, England
JOHN HODGES, Mittersill, Austria
DAVID R. NOTTER, Virginia Polytechnic Institute and State University
DIETER PLASSE, Universidad Central de Venezuela (Retired)
LOUISE LETHOLA SETSHWAELO, Ministry of Agriculture, Botswana
THOMAS E. WAGNER, Ohio University, Athens
JAMES E. WOMACK, Texas A&M University

Board on Agriculture

The National Academy of Sciences is a private, nonprofit, self-perpetuating society of distinguished scholars engaged in scientific and engineering research, dedicated to the furtherance of science and technology and to their use for the general welfare. Upon the authority of the charter granted to it by the Congress in 1863, the Academy has a mandate that requires it to advise the federal government on scientific and technical matters. Dr. Frank Press is president of the National Academy of Sciences.

The National Academy of Engineering was established in 1964, under the charter of the National Academy of Sciences, as a parallel organization of outstanding engineers. It is autonomous in its administration and in the selection of its members, sharing with the National Academy of Sciences the responsibility for advising the federal government. The National Academy of Engineering also sponsors engineering programs aimed at meeting national needs, encourages education and research, and recognizes the superior achievements of engineers. Dr. Robert M. White is president of the National Academy of Engineering.

The Institute of Medicine was established in 1970 by the National Academy of Sciences to secure the services of eminent members of appropriate professions in the examination of policy matters pertaining to the health of the public. The Institute acts under the responsibility given to the National Academy of Sciences by its congressional charter to be an adviser to the federal government and, upon its own initiative, to identify issues of medical care, research, and education. Dr. Kenneth I. Shine is president of the Institute of Medicine.

The National Research Council was organized by the National Academy of Sciences in 1916 to associate the broad community of science and technology with the Academy's purposes of furthering knowledge and advising the federal government. Functioning in accordance with general policies determined by the Academy, the Council has become the principal operating agency of both the National Academy of Sciences and the National Academy of Engineering in providing services to the government, the public, and the scientific and engineering communities. The Council is administered jointly by both Academies and the Institute of Medicine. Dr. Frank Press and Dr. Robert M. White are chairman and vice-chairman, respectively, of the National Research Council.

Preface

Humans have tended sheep, pigs, cattle, chickens, and other domesticated animals as a source of food, tallow, leather, and draft power for more than 10,000 years. It was only as recently as the early twentieth century, however, that deliberate and planned selective breeding began to shift the genetic makeup of domesticated populations in predetermined directions. In some species, divergent goals were set for different populations. For example, cattle were selected and bred for meat or for milk production. In 1905, this division was recognized by the formation of the Dairy Shorthorn Society in Britain.

The application of the science of genetics in this century to the breeding and improvement of livestock has yielded remarkable results. For example, a single dairy cow in the developed world produces nearly twice the amount of milk per year today than her ancestor could produce just 25 years ago. Since 1983, in less than 10 years, pork producers have reduced fat in pigs by 31 percent and saturated fat by 29 percent. In fewer than 50 years, poultry producers in industrialized countries have reduced, by more than half, the number of days required to raise broiler chickens that are ready for market, from 95 days in 1934 to 45 days in 1991.

Feeding and management improvements contribute to these performance achievements in food animals, but central to these developments has been genetic improvement made possible by the availability of diverse breeds and populations that are the sources of genes needed for breed improvement. However, the focus in the agricultural sector has been toward fewer breeds, and breeds that are suited for wide distribution throughout the world. As these new breeds spread, they displace the many breeds and populations that were the genetic sources for their development.

Improvements in animal breeding techniques and production have advanced rapidly. Perhaps even more rapidly during the past 10 years have come the developments in molecular biology and reproductive physiology that will eventually contribute to further improvements in animal breeding and production and to the assurance of a quality food supply. Noting the swift advances in science and technology, the Committee on Managing Global Genetic Resources believes that now is the time to assess the goals and future directions of the livestock research and production communities, with a specific focus on genetic diversity. Improvements in all areas of animal husbandry, including breeding and genetics, reproductive physiology, nutrition, and animal health, can unintentionally contribute to the decline in the number and diversity of breeds and types. Loss of genetic diversity threatens livestock production systems throughout the world. Preserving the capacity to continue to develop modern livestock requires global actions, on national and international levels, to prevent the loss of valuable genetic resources.

During its assessment and deliberations, the committee observed that consensus exists within the livestock production community on the need to develop national and international efforts to conserve and manage livestock genetic resources. The approaches to conservation, however, are markedly different and are generally defined by two philosophies: (1) utilization and (2) preservation. The utilizationist gives the highest priority to using available genetic resources to improve livestock populations. The objective is to increase the rate and efficiency of livestock food and fiber production. The preservationist emphasizes the value of preserving the widest possible spectrum of genetic diversity to meet future needs. Breed preservation is the preferred method of gene conservation.

Reconciliation of these two views is needed, and it must begin with the recognition of their common goal—sustained improvement of livestock for use by an expanding global population. The committee believes that the information in *Managing Global Genetic Resources: Livestock* can provide the foundation for incorporating both views into a cohesive strategy for conservation and management.

This report examines the genetic diversity of the major domestic livestock species. Those used on a global scale include cattle, sheep, goats, pigs, horses, donkeys, buffaloes, chickens, turkeys, ducks, and geese. The report assesses the status of genetic diversity in the most common of these livestock species and the need for actions to conserve them as resources for future agricultural and livestock development.

Although the report focuses on only a few animal species, it notes that many others, for example, camels, llamas, rabbits, and guinea pigs, are important in various countries or regions. In general, the principles and technologies of genetic conservation presented in this report are applicable to all agriculturally important domestic animals.

The committee, established by the National Research Council's Board on Agriculture in November 1986, is concerned with genetic resources of identified economic value. These resources are important to agriculture, forestry, and industry. The committee has been assisted by two subcommittees and several work groups that gathered information or prepared specific reports. One of the subcommittees, chaired by Henry A. Fitzhugh of the International Livestock Center for Africa in Ethiopia, examined the conservation and management of livestock genetic resources and drafted this report. It is one in a series of reports entitled, *Managing Global Genetic Resources.* Previously published reports in the series are *The U.S. National Plant Germplasm System* and *Forest Trees,* both published in 1991. The committee believes that the status of the genetic diversity of commercial fish and shellfish is an important area of study, but the lack of research and documentation in this area precluded it from publishing a separate report.

Soon to be released is the committee's final report, *Agricultural Crop Issues and Policies.* In addition to the examination of crop plants, the committee's main report will address the legal, political, economic, and social issues surrounding global genetic resources management as they relate to agricultural imperatives.

The Subcommittee on Animal Genetic Resources was formed to assess policies and programs for preserving and using animal genetic resources and to identify strategies for meeting present and future needs. Specifically the subcommittee was asked to do the following:

- Examine the uses and status of livestock genetic resources globally.
- Examine methods for using and preserving livestock genetic resources.
- Identify the major problems that limit effective management of global resources.
- Assess the status of international efforts to conserve and exchange animal germplasm.
- Recommend research and development priorities and practical

strategies for animal genetic resources management at national, regional, and global levels.

• Present a global strategy for conserving and managing livestock genetic resources.

Chapter 1 of this report discusses the need to conserve livestock genetic resources, especially in developing countries where many livestock populations have not been sufficiently characterized. It summarizes the utilizationist and preservationist approaches to conservation and emphasizes a position that would be acceptable to these divergent views. Chapter 2 describes the fundamental elements in livestock genetic resource management, including rationale for preserving populations, identifying sampling strategies, and developing preservation methods. Chapter 3 defines the factors that affect genetic variation, and Chapter 4 profiles the biotechnological methods for delineating and manipulating genetic material. Chapter 5 describes examples and organizational elements of national programs that are involved with managing and conserving livestock genetic resources. Chapter 6 presents the committee's recommendations on strengthening international programs for conservation of animal germplasm.

Appendixes include an overview of the major livestock species by total population sizes and global distributions; a discussion of the risks of disease transmission during embryo transfer; and techniques for the collection, handling, and storage of semen.

The issues surrounding the conservation animal genetic resources have been discussed and debated for more than 30 years. Conclusions generally have been similar—diversity in livestock populations is at risk, valid reasons exist to conserve diversity, and organized conservation activities are highly desirable. Nevertheless, active programs to inventory and conserve genetic diversity have been the exception rather than the rule.

Livestock genetic diversity is essential to meet the future needs of global society. Technologies that emerge from scientific research will enhance capacities to conserve and use animal genetic diversity. Ultimately, conservation of livestock genetic resources will require cooperation at the national and global levels. The committee believes that the conclusions and recommendations presented in this report can contribute to improved conservation and management of a major global resource—livestock.

PETER R. DAY, *Chair*
Committee on Managing
Global Genetic Resources:
Agricultural Imperatives

Acknowledgments

Many scientists and policymakers have contributed time, support, and information instrumental to the committee's analyses contained in this report. In particular the committee thanks Keith Gregory, Gordon Dickerson, John Acree, Roger Gerrits, and Henry Shands for their advice and contributions to the study. Others have provided valuable information and insight. These include Eric Bradford, Harvey D. Blackburn, Jerry Callis, Thomas Cartwright, Roy Crawford, Dale Gustafson, William Hohenboken, Kalle Maijala, Ian Mason, Charles Mullenax, Ian Parsonson, Caird Rexroad, Elizabeth Singh, Philip Sponenberg, Jane Wheeler, and Saul Wilson. Many other scientists and policymakers assisted by reviewing selected draft materials and the appendixes. To all of these individuals the committee expresses its gratitude.

Administrative support during various stages of the development of this report was provided by Philomina Mammen, Carole Spalding, Maryann Tully, Joseph Gagnier, and Mary Lou Sutton, and they are gratefully acknowledged. The committee also thanks Joi Brooks and Sherry Showell, interns sponsored by the Midwestern University Consortium for International Activities, for assisting in the development of the report.

Contents

Livestock

Executive Summary

Livestock animals meet a variety of human needs. They are important sources of transport, draft power, fiber, hides, fertilizer, fuel, and nutritional protein in the form of meat, milk, eggs, and processed products, such as cheese. More than half of all protein consumed by humans comes from livestock and fish, which are a more complete source of essential amino acids than are plants. As nations struggle to improve nutrition and to feed their growing populations, demands for the food, fiber, and other products derived from livestock will increase.

As with crop plants, the purpose of livestock genetic resource management is to maintain a reservoir of potentially useful genetic variation culled from the multitude of livestock breeds and varieties within species. Even though there is little danger that the major domestic animal species, such as *Bos taurus* (cattle) or *Gallus domesticus* (chickens), will become extinct, genes from preserved livestock germplasm can be used for developing new types and breeds, adapting existing types and breeds to new conditions, or meeting changing consumer preferences or production requirements.

The application of the science of genetics in this century to breeding and improvement has substantially increased agricultural production, but it could have deleterious effects in the future. Basic to developing new and more productive animals with better nutritional qualities has been the availability of diverse breeding populations that possess genes useful for livestock improvement. As new breeds develop and spread, however, they may displace indigenous breeding populations. Furthermore, as the efficiency of selection within commercial breeds increases, the capacity within a breed for genetic change could become limited. Preserving society's capacity to improve and develop modern livestock requires that actions are taken to prevent the loss of potentially valuable genetic resources. Because future needs cannot be reliably predicted, a broad array of genetic diversity must be conserved.

TWO VIEWS ON LIVESTOCK CONSERVATION

Although a consensus within the livestock community exists on the need to develop national and international efforts to preserve and manage livestock genetic resources, the approaches to conservation differ widely. They hinge on differences in opinion about the importance of traditional livestock breeds, including indigenous breeds, for long-term genetic improvement programs. In general terms, these approaches can be divided into two broad categories: utilization and preservation.

The utilizationist's primary concern is the immediate usefulness of available genetic resources to improve livestock populations. Descendants of animals with documentable, unique biological characteristics are to be maintained for future use. The loss of breeds as distinct identities is not generally a concern, as long as the genes that make these breeds potentially useful are retained in commercial stocks.

The preservationist's primary objective is long-term conservation of genetic resources for future use. This view emphasizes the value of preserving the widest possible spectrum of genetic diversity to be prepared for unpredictable changes or future needs. The greatest possible number of breeds are to be preserved as purebreds.

Differences between the preservationist and utilizationist views largely pertain to perspective and emphasis, but they strongly influence the practices that are favored for managing genetic resources. The committee bases its conclusions and recommendations on the common goal held by proponents of both views—the goal of sustained improvement of livestock for use by an expanding global population. Setting national or global priorities for sustained livestock improvement will depend on a variety of factors, including available funding and technical expertise, national and international needs, and genetic similarity among available breeds and populations. However, the committee believes that at a minimum, conservation should focus on breeds that are (1) of potential economic value, or (2) are both endangered and represent types with unique biological characteristics.

What scientists have come to understand thus far about livestock is impressive. This basic knowledge has been swiftly carried forward by application. With such rapid advances in science and application, it is imperative that policymakers and researchers seize the opportunity now to put into place priorities for the improved preservation and management of livestock genetic resources.

GENETIC DIVERSITY OF LIVESTOCK

Genetic diversity within a livestock species is reflected in the range of types and breeds that exist and in the variation present within each. Examples of types include milk or beef cattle, milk, meat, or wool sheep, and layer hens or broiler chickens. In dairy cattle, examples of breeds include Holstein-Friesian, Jersey, and the Swedish Red-and-White.

For growth and maternal traits in cattle, Cundiff et al. (1986) has shown that differences among breeds substantially exceed those within breeds. This conclusion can probably be extended to other traits and species, and it suggests that genetic variation among breeds is a major component of the readily accessible livestock diversity. Thus losses of unique types and breeds compromise access to their unique genes and gene combinations.

For example, in sheep raised for meat, intensification of production in the early 1970s led to a desire to increase lambing rates in commercial stock. Selection within existing North American breeds could increase the lambing rate by perhaps .02 lambs per ewe annually. Thus achievement of 0.50 in lambs per year (from 1.50 to 2.00 lambs/ewe/year) would have taken 25 years. Yet through utilization of the prolific Finnish Landrace, a breed that was, at the time, little more than a novel indigenous type of Northern Europe with a mean lambing rate in excess of 2.5, a single generation of crossing to current commercial breeds led to a ewe with a lambing rate of about 2.0. This example illustrates the point that within-breed selection can lead to great genetic change, but only over a long period of intense selection. The presence of a breed with the desired characteristics already in place and with the controlling genes already at high frequency greatly enhances the efficiency of the improvement process.

At times, however, within-breed selection will be required to go beyond the limits of existing breeds. The genetic variation within commercial stocks is the ultimate raw material that allows continued improvements or, if necessary, redirection of selection to achieve new results. The recombination and concentration of favorable genes within major commercial stocks under intensive selection have allowed development of animals with productive capacities that greatly exceed those of their ancestors. Likewise, the store of genetic diversity within commercial breeds has permitted relatively prompt responses to changes in societal demands. The change from the relatively fat pigs of the 1950s to the extremely lean pigs of today was achieved largely through within-breed selection.

Historically, the preservation of genetic diversity within commercial breeds has not been an issue. Accurate identification of superior individuals and their intensive propagation have justifiably received much greater emphasis in order to increase production (more milk from a cow, for example). It can be argued that if genetic diversity were declining, the selection of stocks for commercially desired traits would be less productive for developing improved breeds. However, a decreasing rate of response to selection cannot be documented. Yet technological advances in animal breeding and reproductive control have the potential to change this situation. Embryo transfer and cloning, molecular aids to selection, and greater control of reproductive processes can produce rapid and considerable increases in both the accuracy of selection and in the ability to propagate selected individuals. Through the use of follicle maturation techniques, in vitro fertilization, and embryo cloning, a pair of parents could produce hundreds, or even thousands, of progeny. Given this prospect, consideration of the maintenance of genetic diversity must become part of livestock improvement programs.

Origins of Traditional Breeds

Hoofed animals, specifically cattle, goats, pigs, and sheep, first began to be domesticated about 11,000 years ago. They quickly became essential to the establishment and spread of agriculture. A wide array of subpopulations evolved through adaptation to different environments, as humans migrated, over several thousand years and through breeding efforts during the past 200 years. These subpopulations are generally recognized as breeds, that is, groups of individuals that are similar in their appearance and characteristics. Their genetic diversity has been shaped by the need to adapt to temperature extremes, humidity, variations in quality and availability of feed supplies, diseases and parasites, and other factors that are regionally and globally divergent. They represent easily recognizable genetic resources that can serve as a convenient and reliable source of genes associated with particular physiological or morphological characteristics.

For example, Criollo cattle in South America are descended from Iberian stock brought to the Americas during the 1500s. Over the past 500 years they have developed traits, through natural selection, that enable them to live on poor nutritional levels and to withstand climatic extremes. Genetic diversity has allowed for their survival, and they are now a source of genes that may contribute to adapting other breeds for production in comparable environments around the world.

One of the challenges in conserving genetic resources is to prevent excessive losses of unique, adapted breeds. They need to be maintained, and their potential contribution to future commercial stocks should be carefully assessed.

Displacement of Traditional Breeds

The traditional indigenous breeds, particularly in developing countries, are generally considered to be at greatest risk of genetic loss. In the tropics, temperate breeds usually cannot be introduced as purebred animals without also introducing high levels of feeding, special management, investments, and risks. Small farmers often cannot afford the cost of the high-maintenance breeds that are being introduced. However, these breeds are commonly crossbred with local animals to combine adaptation and disease resistance with higher production. Although the traditional indigenous breeds are needed initially to produce crossbreeds, and perhaps over the longer term to create a stable self-perpetuating subpopulation, their critical role in maintaining self-perpetuating breeding programs is often not recognized. Farmers may be reluctant to keep the indigenous breeds if they are less profitable than crossbreeds. As a result, indigenous breeds lose their genetic identity or disappear.

Although the extent and usefulness of genetic variation among most traditional breeds have not yet been fully characterized, the differentiation clearly reflects a wealth of genetic diversity within the species—diversity that permits further selection and improvement. In particular, these populations hold genetic combinations that may prove valuable in meeting future environmental, production, or marketing conditions.

In many temperate regions, a number of local breeds have already been replaced by high-performance and more economically viable breeds. For example, in Great Britain, the English Longhorn was the dominant cattle breed during the nineteenth century. It was prized for its size and fat, because draft work and tallow for candles were the important products of the time.

Beginning only about 200 years ago, when the horse became the preferred draft animal, the enclosure of common land made controlled breeding possible (Mason, 1984). Cattle feeding practices also improved (Mason, 1984). By the turn of the century, selection for milk and beef production became paramount and the Longhorn's position gave way to the Shorthorn, a dual-purpose, milk and beef breed. By the mid-twentieth century, the Holstein-Friesian, a specialized dairy breed, was on the rise in the United Kingdom, and large continental breeds, such as the Charolais, were being used as sire lines for beef production. The flexibility to change genotype relatively quickly to meet changing market requirements results from the genetic variation between breeds. This same flexibility will likely be required under future, unforeseen production conditions.

The Impact of Technology

Recent biotechnological advances can further alter genetic structure and limit the diversity of livestock species. Cryopreservation of semen and embryos (storage typically in liquid nitrogen at a temperature of –196°C) is now feasible for many livestock species. Germplasm can be stored indefinitely and shipped worldwide. Health concerns, traditionally a barrier to international transport of livestock, may soon be resolved by special handling techniques for embryos that provide better control against the spread of pathogenic organisms. The international transport of germplasm can be expected to increase. Concurrently, technology advances for storing and transporting semen and embryos have allowed the trend toward widespread use of germplasm from a small number of individuals, which could cause overall genetic diversity to decline.

Regardless of geographic location, when production and marketing systems become more uniform or when the influx of foreign germplasm of one breed is allowed to dominate the population, the danger of narrowing the genetic base of livestock species exists. The application of new technology for germplasm propagation and dissemination may contribute to the erosion of diversity. Unique germplasm is threatened by replacement of breeds with more productive or popular stocks, dilution of breeds through crossbreeding programs, and decreased diversity within highly selected breeds or lines that have a small number of breeding individuals. The committee believes that it is now time, scientifically and socially, to address these concerns in the design of programs for livestock improvement.

LIVESTOCK PRODUCTION SYSTEMS

Production and management systems for domestic livestock can be broadly classified as either intensive or extensive. Intensive sys-

A young female commune worker in the People's Republic of China collects eggs from chickens kept in cages, a common intensive production approach used for poultry. Credit: Dean Conger, ©1992 National Geographic Society.

tems were developed in the latter half of the twentieth century and involve high levels of inputs, such as supplemental feed and disease prevention efforts, and require more capital investment, such as housing construction and equipment. Chickens, for example, are now commonly raised in confined, environmentally controlled housing with sophisticated disease control programs. This type of system is more frequently found in developed countries, which have the capital to invest, plentiful grain supplies for animal feed, and supportive technologies such as veterinary care and modern production facilities.

The stocks of chickens, dairy cattle, pigs, and turkeys that now predominate in developed countries have been selected for high production under intensive management and optimal environmental conditions. The genetic base of these specialized stocks may become more narrow because of intensification of selection and smaller effective population sizes (the number of reproducing individuals in a population). Examples of specialized stocks are Leghorn chickens, which are superior for egg production, and the familiar black and white Holstein-Friesian cattle, which dominate other dairy cattle breeds because of higher milk production.

The term *extensive* describes the more traditional systems used to raise livestock. Fewer purchased inputs, such as high-quality feed or pharmaceuticals, are provided. For ruminants, this approach involves maintaining animals, such as cattle, goats, and sheep, under range conditions. Backyard poultry and swine production is another example of extensive systems. Nutritional supplementation and veterinary care are minimal or nonexistent.

In extensive systems, animals are generally less protected from a variety of environmental and nutritional stresses, and they can become adapted to local conditions over long periods of time (several generations). The N'Dama cattle of West Africa, for example, are genetically tolerant of the effects of trypanosome infection and, thus, can survive in an area that is inhospitable to susceptible breeds. In Africa, about 30 percent of the 147 million cattle in countries inhabited by the tsetse fly, which transmits trypanosomiasis, are exposed to the disease. The disease can cause poor growth, weight loss, low milk yield, reduced capacity for work, infertility, abortion, and early death. Annual losses in the value of meat production alone are estimated at $5 billion. Additional costs come from losses in milk yields, tractive power, waste products that provide natural fuel and fertilizer, and secondary products, such as hides (International Laboratory for Research on Animal Diseases, 1991).

Animals in extensive production systems, like this flock of sheep in Tunisia, usually forage and water on open pasture. Credit: Food and Agriculture Organization of the United Nations.

WHY CONSERVE THE GENETIC DIVERSITY OF LIVESTOCK?

Extinction is not a concern for the major domestic species, which benefit from society's protective custody. Large total population sizes are maintained to ensure adequate product supplies, and the species' basic needs are usually met. Total population sizes are generally in the hundreds of millions, even though the effective population size may be much smaller.

The compelling need for conserving domestic species is to prevent the loss of the many differentiated populations that, because of geographic or reproductive isolation, have evolved distinct characteristics and now occupy different environmental niches. A further need exists to guard against depleting diversity due to current breeding technologies, worldwide movement of germplasm, the proliferation of highly selected industrialized breeds, and commercial stocks founded on a relatively small number of breeding individuals. The range of genetic diversity in livestock species must be saved as a foundation for future improvements and adjustments to changing production conditions.

Because the ultimate effects of diminished genetic diversity are hard to estimate, given the unpredictability of future needs, the conservation of livestock genetic diversity can best be considered as a form of insurance. Preserved stocks will possess potential economic, scientific, and sociocultural benefits.

Economic Benefits

Genetic variation provides the foundation to continue improvements in production efficiency and to accommodate, relatively rapidly, unpredicted changes in the methods for producing animals and marketing their products, or in the demand for animal products. An example of the need to respond to changing demands is seen in the pork industry where several decades ago animals that possessed large amounts of fat for lard were desired. With the advent of consumer concerns about animal fats, and the ready availability of vegetable shortenings, the demand for lard declined. Consequently, breeders produced the leaner, meat-producing animals prevalent today.

Scientific Benefits

Studies of genetic variation and gene variants provide insight into physiological and biochemical functions, gene structure and control, and evolutionary processes in animals. Miniature pigs (a genotype commonly used in studies of growth, obesity, and diabetes) illustrate the use of a domestic stock to further scientific investigations. Such studies are beneficial to both animal and human health.

Sociocultural Benefits

Genetic diversity may be of social or cultural value. Some animals are used for recreation, and animal breed exhibits and fairs are attractions for many people. Genuine cultural benefits can be realized through the use of animals to visually chart important historical developments—certain breeds are living evidence of a nation's agricultural heritage. In Hungary and France, for example, traditional indigenous breeds are maintained in national park settings and serve as tourist attractions. Traditional livestock breeds are essential components of living historical farms in Europe and North America. In addition, specific livestock breeds characterize the way of life and provide specific products for many villages and regions around the world.

TECHNOLOGIES FOR CONSERVING AND USING GERMPLASM

Recent technological developments may, in some instances, have a negative impact on genetic diversity by reducing the number of parents required to propagate a population. However, they also open important avenues for expanding the scope and assessing the potential value of conservation efforts. Improved reproductive technologies enhance the feasibility of preserving germplasm through better reproduction rates and disease control. Molecular technologies provide more understanding and control at the level of the gene, and should vastly improve opportunities to use preserved stocks.

Molecular Technologies

New molecular genetic technologies will improve the use of preserved stocks and aid in assessing genetic distance and variation. For example, individual animals could be selected for breeding and preservation based on specific knowledge of their genotypes. Molecular studies may provide greatly enhanced understanding of the genetic basis of important livestock production traits. These technologies also promise the capacity to move routinely important genes between unrelated animal species.

Characterization of Genomes

Methods to characterize the genomes of individual animals and the gene pools of breeds are becoming a reality. A variety of techniques is available to quantify genetic variation at the levels of the gene product and of the gene itself. Numerous studies have examined variation in the protein products of the genes, but as molecular technologies become more sophisticated, the focus is shifting to the genes and the DNA (deoxyribonucleic acid) to quantify differences among individuals, breeds, and species. Genetic distances, indexes of the relative genetic differences between two or more breeds, can be estimated. This information can be used to aid decisions regarding preservation that are based on the genetic uniqueness of a germplasm source.

Gene Transfer and Mapping Technologies

The expansion of research on the effects of specific genes, and on their transfer between individuals of the same or different species, has major implications for the potential value of preserved germplasm. It should be possible to screen preserved stocks and, if desired, to extract and use a beneficial gene or genes that may be located in an otherwise undesirable background. For example, the International Livestock Center for Africa, in Ethiopia, and the International Laboratory for Research on Animal Diseases, in Kenya, are cooperatively conducting research to identify the genetic basis of tolerance to trypanosome infection. This knowledge (and perhaps the genes themselves) may expand the tolerance to trypanosomiasis of livestock across breeds and even species.

Efforts to map livestock genomes are progressing. Advances in sequencing and mapping techniques that emerge during this effort will enhance, and be enhanced by, identification and use of human genes.

Genomic Libraries

Today's technology permits the formation of genomic libraries theoretically containing the full set of genes from any given individual. However, a tremendous amount of research must still be done to identify genes of importance. Regeneration of living animals from DNA stores in genomic libraries is not yet possible, and genomic libraries should only be considered as an adjunct means of germplasm storage at this time. Their potential value as a preservation tool, however, can be likened to a form of insurance. It is anticipated that methods will improve for isolating and manipulating specific DNA segments.

Animal Reproduction Technologies

Animal reproduction technologies include multiple ovulation, in vitro fertilization, embryo transfer, embryo splitting, cloning, sexing, and cryopreservation. The cryopreservation of gametes and embryos is a well-established technology in livestock production, although successful application to a wider array of livestock species and to oocytes and embryos requires more research and development. Notably, methods to cryopreserve pig embryos have been unreliable. Although widespread use of these technologies can narrow the breeding population and thereby have negative impacts on genetic diversity, these technologies can also, if properly used, become tools for support of genetic resource management.

An important development in embryo research has been the use of hormones that induce multiple ovulations in a donor female and lead to a greater number of embryos. This capacity is especially important in species that normally produce only one offspring per pregnancy. Further, embryos can be cloned to produce genetically identical animals, providing much greater flexibility in distribution and research efforts.

One of the main limitations to the movement and use of genetic resources has been concern about the health status of the germplasm. If, in the course of importing livestock, exotic diseases are inadvertently introduced into a region, the effects on indigenous stocks could be devastating. Most nations have not been willing to take the risks involved in importing livestock. However, methods for the screening, diagnosis, and control of diseases are improving, and new embryo-washing techniques offer promise for eliminating disease agents. The international movement of animal breeding stock as frozen semen and embryos is thus becoming more feasible.

METHODS OF PRESERVING ANIMAL GERMPLASM

Three basic approaches can be identified for preserving genetic diversity: maintaining living herds or flocks, cryopreserving gametes or embryos, and establishing genomic libraries. The vast majority of livestock genetic resources will continue to be maintained in living herds and flocks, many of which are privately owned. If the size of the breeding population is sufficiently large and the population is not decreasing in number, directed conservation efforts will not be required. Periodic inventories of animal populations can provide early warnings of any changing patterns in production or use that may alter the diversity within breeds. A major advantage of preserving live herds and flocks is the opportunity for selection, thereby allowing the breed to adapt to shifting environmental conditions.

Frozen storage of gametes and embryos offers a cost-effective method to preserve the genetic material of a breed for an indefinite period of time. Collection and freezing of semen is relatively simple and, if samples are collected from a sufficient number of males, allows the preservation of essentially all the genetic variation in a stock. The costs of collection are not excessive, although in remote areas where accessibility to equipment and facilities may be problematic, they can be higher. The cost of maintaining the samples in frozen storage is low. Methods for collecting, handling, and freezing embryos, although more complex than those for semen, also offer efficient means of preservation. Embryo storage has not been used for conservation, in part because of the cost of sampling, but research on collecting and handling oocytes and embryos is advancing rapidly. Cryopreservation can complement efforts to preserve live populations, and it should be used as a safeguard when population numbers are dangerously low or when certain breeds or lines are likely to be replaced with other populations.

As noted earlier, genomic libraries are of little use for breed preservation. As technologies develop, however, they may provide an important mechanism not only for conserving diversity, but also for accessing particular genes. Their value will be enhanced with continuing advances in molecular genetics research.

Criteria and priorities must be established to identify which genetic resources merit conservation. Considerable attention has been given to determining the basis for classifying a breed as endangered, and guidelines for various species are available to aid in evaluating preservation needs (Maijala et al., 1984). However, the status of each breed must be evaluated on a case-by-case basis. In general, an endangered breed is defined as having a small and decreasing number of breeding individuals. Unfortunately, in developing countries, frequently little or no information is available on certain breeds, and information cannot be quickly gathered. At times, extemporaneous judgments must be made about which individuals constitute a breeding population.

Once a decision is made to preserve a breed, a sampling strategy must be developed. Factors such as geographic distribution and population structure, if known, should be taken into consideration. A stratified system of sampling (that is, identifying breeds and taking samples of unrelated individuals and at random) will in most situations be appropriate.

The effective preservation and management of livestock genetic resources will require organization and accessibility of basic information on animal breeds in a data base. Although some descriptive and analytic information exists for many livestock breeds, it still may be insufficient for making management decisions. Studies to inventory, characterize, and compare livestock populations are badly needed, particularly in developing countries. The cataloging of information so it can be readily retrieved will be critical. Information accruing from research on gene mapping, gene transfer, and related technologies must also be cataloged in an accessible format.

EFFORTS TO IMPLEMENT LIVESTOCK CONSERVATION

The issues surrounding the conservation of animal genetic resources have been discussed and debated for more than 30 years. Conclusions generally have been similar:

- Diversity in livestock breeds is increasingly at risk,
- Valid reasons exist to conserve diversity, and
- Organized preservation and management activities are highly desirable.

Nevertheless, active programs to inventory and preserve genetic diversity are the exception rather than the rule. This situation appears to be changing as more developed and several developing countries implement preservation and management programs for animal genetic resources.

The lead on discussions of genetic resource management has largely been taken by the Food and Agriculture Organization (FAO) of the United Nations, with support from the United Nations Environment Program (UNEP). The FAO has contributed through its comprehensive publications on characterization and management techniques of animal genetic resources. It has also recognized the need for collating information contained in central data bases, and has supported studies to determine how best to establish and organize such data bases. In 1989, the Commission on Plant Genetic Resources of the FAO voted to address conservation of animal genetic resources in addition to plants, thereby demonstrating its recognition of the importance of livestock genetic resources in global agricultural production.

In 1992, FAO announced the initiation of a program to conserve and develop the livestock and poultry genetic resources of developing nations (Henson, 1992; Ruane, 1992). Its five elements are a global inventory, breed preservation, indigenous breed development and conservation, gene technologies, and a legal and international framework. The Global Data Bank for Domestic Livestock, based in Rome,

Italy, covers all areas of the world with the exception of Europe, but including the Commonwealth of Independent States (the former Soviet Union). This data base will document and characterize the different populations of domestic livestock.

The loss of local minor breeds has been a concern in Europe for many decades. The European Association for Animal Production (EAAP), based in Italy, formed a group to coordinate activities relating to livestock resources of Europe. National conservation programs have been established both in Europe and in several developing countries. The FAO and EAAP have promoted establishment of a regional data base for Europe. The Animal Genetic Data Bank at the Hanover School of Veterinary Medicine in Germany records census data on populations of breeds and genetic characterization data. The data bank will be valuable to educational institutions, governments, commercial interests, and others who are concerned with the conservation of animal genetic resources.

RECOMMENDATIONS

Livestock genetic diversity is an essential component of livestock improvement programs, especially to meet the future needs of human society. There is a paucity of quantitative evidence to support arguments for the need to conserve livestock genetic resources. However, when examining the use made of landraces in plant improvement in both developed and developing nations, the committee concluded that it is prudent to preserve unique sources of germplasm in animals as well.

Ensuring that livestock resources will be preserved and properly used will require expansion and coordination of existing efforts and implementation of new ones. Technologies that emerge from scientific research will enhance the present capacities to preserve and use animal genetic diversity. Ultimately, conservation of livestock genetic resources will require cooperation at the national and global levels. The following are the committee's major recommendations for addressing the management and use of livestock genetic resources. Additional recommendations and further detail are presented in the main text.

Expanded Programs and Activities

Mechanisms must be put in place to ensure that genetic diversity of the major livestock species is maintained to support improvements in production efficiency and to accommodate future changes in selection goals.

The issues surrounding the need to conserve livestock resources have been discussed for many years in a variety of forums. For example, FAO forums have provided a wealth of guidelines for establishing programs, and the FAO and UNEP have worked together to supply training and to organize regional meetings on animal genetic resources. Additional activities have been ongoing in Africa, Asia, Europe, and Latin America. All of these efforts recognize the urgent need to integrate conservation and livestock improvement efforts. If action is not taken soon, potentially valuable livestock genetic resources could be lost.

Documentation of Breeds

Increased efforts are needed to inventory and characterize unique and endangered populations and breeds, particularly in developing countries.

Decisions about needs and priorities in livestock conservation programs must be based on knowledge of the animal populations. For populations found in developing regions, the need for documentation is particularly acute, because industrialization and the importation of improved breeds can threaten indigenous breeds. An essential component of this process is establishing national and international data bases that maintain information in a readily accessible form. The Animal Genetic Data Bank at the Hanover School of Veterinary Medicine and FAO's world inventory of native livestock breeds and strains constitute a good beginning, but much greater effort is needed to collect information, particularly from developing countries.

Increased Cooperation

Private efforts to preserve animals in breeding herds should be encouraged.

Private farm parks, such as those associated with the Rare Breeds Survival Trust in England, have proved successful in preserving selected rare or historic breeds. In some cases, governments have subsidized or otherwise assisted farmers who elected to maintain animals of particular breeds. In Canada, for example, farmers in the Province of Quebec are paid for rearing purebred Canadienne cattle. Cooperation also could take the form of exchanging information about the breeds preserved or of providing technical assistance on breeding programs to control inbreeding or ensure breed purity.

Although the many herds and flocks held by private individuals and groups may not originally have been assembled for conservation motives, these populations can supplement public activities. The efforts of these groups enable a broader diversity of breeds and populations to be preserved than would otherwise be possible with national resources that are frequently limited. Responsible private efforts to preserve livestock germplasm should be encouraged and offered public support.

Research and Technology Development

Research on technologies that could benefit preservation and use of animal genetic resources should be continued and expanded.

Research in new technologies, such as genome mapping and gene transfer, is moving at a rapid pace. Future breakthroughs are likely to have significant effects on the efficiency of animal production and on the ability and need to preserve germplasm. Major research efforts are under way on the human genome, and complementary work is progressing on the genomes of cattle, pigs, and sheep. Preliminary results suggest that much of the genomic information from one species will be applicable to others. To maximize the efficiency of basic and applied research, geneticists working in different areas and on different species must take the initiative to cooperate and share information. A number of research areas have potential to enhance the preservation, management, and use of animal genetic resources.

Cryopreservation as a Supplement to Breeding Populations

The preservation and management of endangered and unique populations as breeding herds or flocks should be supplemented by cryopreservation of their germplasm.

Cryopreservation of semen and embryos complements preservation of live populations and provides a safeguard when population numbers are dangerously low or when breeds or lines are likely to be replaced or lost. It offers the potential to store large numbers of genotypes for indefinite periods of time. In general, the costs of collecting and processing samples for cryopreservation are not excessive for domestic animals, particularly in regard to semen.

At present, semen from the major livestock species can be cryopreserved, but the process is only reliable enough to be routine and commercial for species such as cattle and, increasingly, sheep. For pigs, goats, horses, and poultry, cryopreservation methods, although available, need further refinement. The effectiveness of cryopreservation for preserving species and breeds of minor livestock is less well established. With semen storage, it is not possible to derive purebreeding stock from stored germplasm unless suitable female recipients are available. However, the most common future use of cryopreserved germplasm is expected to involve crosses with commercial populations to introduce new genetic traits and develop new breeds. Frozen semen is ideal for such use.

Mouse embryos were reported to be successfully cryopreserved, thawed, and implanted in 1972. Since then, the procedures have become routine in several mammalian species, including cattle, sheep, rabbits, and mice. Notably, efforts to cryopreserve pig embryos have been unsuccessful. Cryopreservation of male and female embryos preserves the entire genome of a breed and is particularly appropriate when regeneration of purebred animals is desired.

National Initiatives

National programs for preserving, managing, and using livestock genetic resources must be established in developed and developing countries and suited to different needs.

Before global efforts can be effective, conservation of animal genetic resources must be a national initiative. Given the substantial differences in needs and appropriate approaches in individual countries, the organization of national conservation programs will vary widely. It is important that private and public interests within each country cooperate in developing a strategy to ensure an effective, coordinated national program. The support of a strong network of national programs is essential to the success of international conservation efforts.

Mechanisms for providing bilateral and multilateral support to enable developing nations to preserve, manage, and use their livestock genetic resources must be developed.

Multilateral and bilateral aid programs will be required in many cases to help establish national preservation and utilization efforts in developing countries. Technical expertise from the developed countries should also be readily available. Conservation of genetic diversity is a global concern, and given the greater threat to uncharacterized, unimproved livestock populations in developing countries, support from conservation programs by developed nations is essential. Ultimately, livestock production and improvement programs of both developed and developing countries will benefit from international support and assistance and from enhanced international movement of germplasm resources.

A Global Organization

A global mechanism should be established to provide leadership and support to nations to ensure the adequate conservation of livestock genetic resources.

International leadership is needed to coordinate and facilitate national and regional efforts, and to foster cooperation among national efforts. An organization with leadership responsibilities could assemble experienced specialists to develop priorities and guidelines for global cooperation. This organization must have the confidence of financial donors in its ability to succeed, scientific capability, administrative expertise, and a consultative mechanism.

The committee examined three potential options for achieving the necessary global leadership and coordination:

- Establishing an institute within the Consultative Group on International Agricultural Research (CGIAR);
- Expanding the efforts of the FAO; and
- Expanding support to an existing institution, such as the World Conservation Union (IUCN), formerly the International Union for Conservation of Nature and Natural Resources.

Organizing a new program within the CGIAR or expanding the existing International Board for Plant Genetic Resources (IBPGR) would require new financial resources and considerable study. The FAO's work in this area spans more than 4 decades, and it has an experienced administration in place that could manage a conservation program. However, it has certain limitations. For example, although FAO has a clear mandate to fully develop many of the actions recommended by its advisory panels, its funding is inadequate.

Most conservation-related organizations are not structured to deal with agricultural germplasm preservation. Some reorientation of programs, funding, philosophy, and purpose may be required. An organization that might undertake responsibility for animal genetic resources is IUCN, which has been involved in conservation activities for more than 40 years. It has experience in developing data bases for supporting conservation activities with wildlife, monitoring natural areas, and fostering international cooperation for conservation (Office of Technology Assessment, 1987). Expansion of its mandate would be needed to include livestock genetic resources, and philosophical conflicts between the needs of agriculture and the organization's environmental priorities are likely.

Given the need to establish global leadership in guiding, coordinating, promoting, and instituting animal genetic resource conservation efforts, the committee considers the FAO to be an appropriate institution for these efforts. However, FAO should call on the expertise of other national and international institutions to achieve its goals.

In conclusion, a broad base of genetic diversity in domestic animal species must be maintained to provide flexibility to improve livestock for today's production conditions and to meet unforeseen future production requirements. Technological advances in identifying and propagating superior types may allow for substantial losses in genetic diversity. These situations suggest that designers of livestock improvement programs must begin conscientiously to consider their potential impact on genetic diversity.

1

The Need to Conserve Livestock Genetic Resources

Selected varieties of plants and animals form the foundation of agriculture. Domesticated animals have provided people with food, fiber, transport, and draft power for thousands of years, and during the past 200 years, they have been modified by selective breeding to serve better the changing needs of humanity. Genetic diversity within agricultural species is important as expanding human populations seek to improve their standards of living and place greater demands on natural resources. New technologies that manipulate animal physiology and environment have substantially enhanced livestock production, particularly in the latter half of this century, to meet food and other economic demands.

As a result of modern changes and improvements, however, the genetic diversity of some species may be declining. Symptomatic of that decline is the increased reliance on animals of a single breed or type for commercial production at the expense of other recognized types. If genetic diversity is significantly reduced, it may limit future options for improving livestock populations or for modifying them to meet unforeseeable needs.

THE DEVELOPMENT OF ANIMAL AGRICULTURE

Domestication of livestock began about 11,000 years ago, following the rise of grain cultivation (Hodges, 1990a; Willham, 1982). As the security of plant-based food supplies increased, people adopted a more settled life-style in small communities. However, decreased mobility intensified hunting around the settlements and nearby animals became scarce. Domesticated animals offered a dependable source of meat, skins, and bones; later, their value for other products and uses was recognized. Ruminants, such as cattle, goats, and sheep, were particularly beneficial because they ate forage plants that were unsuitable for human consumption.

Goats and sheep were the first species to be domesticated for food, beginning about 11,000 years ago. Domestication of pigs and cattle followed and was complete by about 8,000 years ago. The earliest archeological traces of these domesticates are found in the Fertile Crescent, an area of southwest Asia between the Mediterranean Sea and the Persian Gulf. Horses were probably tamed about 6,400 years ago in central Eurasia; chickens were domesticated about 5,000 years ago in Southeast Asia; and buffalo were domesticated about 4,000 years ago in India and Southwest Asia. In the Americas, there is evidence of the taming of alpacas and llamas in the Andes Mountains up to 6,000 years ago and the domestication of turkeys in Central and South America about 2,000 years ago. (Reviews of the domestication of these and other animal species can be found in Clutton-Brock [1981], Epstein [1969, 1971], Mason [1984], and Zeuner [1963].)

About 40 domesticated mammalian and avian species are widely recognized (Table 1-1). However, only 6 mammalian species (buffalo, cattle, goats, horses, pigs, and sheep) and 4 avian species (chickens, ducks, geese, and turkeys) are used extensively throughout the world because cultural preferences and use for them have developed. These species are classified as the major domesticates. (See Appendix A for an overview of the status of these species.)

Although the other minor species are numerically less abundant on a global scale, they are nevertheless critically important to the people whose livelihoods are built around them (National Research Council, 1991). The uses of the minor species depend primarily on their abilities to produce valued products in specific environmental niches and on the traditional or cultural patterns of use for their products in some societies. Species have developed marked physiological distinctions under different environmental conditions, and accordingly, some breeds are far better suited than others for production under specific conditions. Examples are guinea pigs, alpacas, llamas, and yaks in high altitudes; camels in desert regions; and musk oxen and reindeer in arctic tundra.

In addition to genetic variation among species, genetic differences are also found between types and breeds of a single species. For example, hair sheep are recognized for their superior ability to

TABLE 1-1 Generic Distribution of Important Domesticated and Semidomesticated Mammals and Birds, by Order, Family, and Species

Mammals	Birds
Lagomorpha	Anseriformes
Leporidae	Anatidae
Oryctlagus cuniculus (rabbit)	*Anas platyrhynchos* (duck)
Rodentia	*Cairina moschata* (muscovy duck)
Muridae	*Anser anser* (goose)
Rattus norvegicus (rat)	*Branta canadensis* (Canada goose)
Mus musculus (mouse)	Galliformes
Caviidae	Phasianidae
Cavia porcellus (guinea pig)	*Gallus gallus* (chicken)
Carnivora	*Coturnix coturnix* (Japanese quail)
Canidae	*Phasianus colchicus* (ring-necked pheasant)
Canis familiaris (dog)	*Pavo cristatus* (peafowl)
Felidae	Numididae
Felis catus (cat)	*Numida meleagris* (guinea fowl)
Proboscidae	Meleagrididae
Elephantidae	*Meleagris gallopavo* (turkey)
Elephas maximus (Asiatic elephant)	Columbiformes
Perissodactyla	Columbidae
Equidae	*Columba livia* (pigeon)
Equus caballus (horse)	
E. asinus (ass or donkey)	
Suidae	
Sus domesticus (swine)	
Camelidae	
Camelus bactrianus (Bactrian camel)	
C. dromedarius (Arabian camel)	
Lama pacos (alpaca)	
L. glama (llama)	
Cervidae	
Rangifer tarandus (reindeer)	
Bovidae	
Bos taurus (humpless, primarily temperate zones)	
B. indicus (humped, primarily tropical zones)	
B. grunniens (yak)	
Bibos sondaicus (banteng)	
B. frontalis (gayal)	
Bubalus bubalis (water buffalo)	
Ovibos moschatus (musk ox)	
Ovis aries (sheep)	
Capra hircus (goat)	

NOTE: Not covered in this report but also of value are certain reptiles (for example, crocodiles, turtles), insects (for example, silkworms, honeybees), fish and shellfish (for example, carp, salmon, crayfish), doves, red deer, and fur-bearing mammals.

produce meat and milk in tropical and subtropical climates, while wool sheep are more suited to producing fiber needed by people in cooler regions. In cattle, the *Bos indicus* breeds from the Indian subcontinent are generally better adapted to hot, humid climates than are the European *Bos taurus* breeds, which, although genetically su-

A PROBLEM OF NOMENCLATURE

Carolus Linnaeus, a Swedish botanist during the 1700s, developed a system of naming and classifying plants and animals. According to his system, cattle belong to the order Artiodactyla or even-toed hoofed mammals (ungulates), the family Bovidae, and the genus *Bos*. Domesticated cattle are usually classified into either the humped (*Bos indicus*) or humpless (*Bos taurus*) types. Most of the humpless cattle include ancestors of the European cattle and the majority of the cattle, which came from Europe, found in the temperate regions of North and South America. The humped cattle, usually called zebu, originated in Paki-

N'dama cattle (left page) are a humpless breed, Bos taurus, *noted for relatively high production of meat and milk per unit body weight and its resistance to trypanosomiasis and other endemic diseases of West Africa. Credit: International Livestock Center for Africa. Zebu (right page) from*

perior for milk production and growth rate under temperate conditions, usually fail to produce adequately or to survive under tropical conditions (Cunningham and Syrstad, 1987). However, some *Bos taurus* breeds that were established in West Africa and Latin America before the introduction of the *Bos indicus* have successfully adapted.

---

stan and India and later spread into the tropical zones of Asia, Africa, Australia, Central and South America, the Caribbean, and the southern United States.

The designation of cattle as two species appears to be inappropriate because *B. indicus* and *B. taurus* have the same chromosome number and interbreed readily; the crosses between them are fertile. However, because of historic origins, the types continue to be distinguished by their original species names. They also represent types that are widely divergent in their adaptation to various environments. The information that will come from genetic maps of *B. indicus* and *B. taurus* may one day help determine how closely related the various breeds are within these species.

Chad in equatorial Africa are Bos indicus *cattle with pronounced humps and fleshy dewlaps that aid heat regulation in hot, humid climates. Credit: Food and Agriculture Organization of the United Nations.*

THE EFFECTS OF TECHNOLOGY ON GENETIC DIVERSITY

Technological advances, such as artificial insemination, and changing practices, such as better disease control, in all areas of animal husbandry, including breeding and genetics, reproductive physiology, and nutrition, are contributing to the improved production capacities of modern livestock. However, these and other techniques have the potential to affect negatively the number and diversity of breeds and types in both developing and developed regions. Such techniques include in vitro fertilization and cryopreservation of semen (storage typically in liquid nitrogen at a temperature of –196°C).

Many livestock species traditionally have been raised under extensive farming conditions, that is, on the range or in fenced areas, often with minimal inputs for housing, labor, nutritional supplements, and veterinary care. Their survival has been challenged by diseases, parasites, and nutritional and climatic stresses over many centuries, allowing time for natural selection to create populations genetically adapted to local conditions.

In developing countries, where livestock management practices and inputs have changed little, adaptation to the environment re-

Many village families in Botswana keep free-ranging pigs and poultry. These animals do not receive veterinary care and are fed only household scraps, but they provide meat and other products for the family and cash from sales. Credit: Elizabeth Henson, Cotswold Farm Park.

mains critical. The N'Dama cattle of West Africa, for example, have developed specific resistance to trypanosomiasis, a parasitic disease spread by tsetse flies (Trail et al., 1989). The cattle survive in an environment in which other, more recently introduced, breeds become unproductive and often die. Yet tremendous pressure to increase food production exists in these developing countries, and the highly productive breeds of livestock developed in the industrialized world may superficially appear ideal for use. The technology for sending germplasm to Africa exists, but the capacity to replicate the improved environmental conditions of the industrialized world does not. The result can be the introduction of unadapted livestock that survive poorly and do little to improve food production.

Inputs and management technologies, such as climate control, are more common in developed countries and reduce the need for livestock to be adapted to the environment. Yet a breed's capacity to be adapted to new conditions, which requires genetic diversity, remains critical to meet future needs and to respond to future production changes.

For example, beginning in the 1940s in developed countries, cages were adopted by chicken farmers to have cleaner eggs and to reduce cannibalism and feather pecking. The cage system was also a better alternative to land that could harbor the disease organisms contained in fowl droppings. Production and management advantages, such as automated feeding, watering, egg collection, and better feed conversion efficiency, led to further adaptation. However, a move to ban the use of cages in the European egg production industry was begun in the early 1980s in response to animal welfare concerns. The Council of Ministers of the European Community has established space and other minimum requirements for hens that will apply to all cages as of January 1, 1995; Switzerland has banned cages; and Sweden and the Netherlands may also ban cages (U.S. Department of Agriculture, 1991). These changes may now require the development of poultry breeds better suited to cageless environments.

Breeding and Improvement Technologies

Increased production of milk, meat, eggs, and animal fiber has been based on extensive records of pedigree and performance, which are essential for accurately identifying superior individual animals. Systematic improvement of livestock populations began in eighteenth-century Europe. Livestock were bred, and the best animals were selected for breeding until a degree of uniformity was attained. Herdbooks were started to record pedigrees because of the reliance on ancestry.

The science of population genetics, developed in the 1930s and 1940s by scientists in Europe and North America, has since been effectively applied to achieve substantial increases in animal production and productivity, primarily in developed countries. Progress has been greatest in dairy cattle, pigs, and poultry. Compared with 30 years ago, the average dairy cow in the United States produces over twice as much milk (U.S. Department of Agriculture, 1976, 1990), and fat thickness in Danish pigs had declined by half (Christensen et al., 1986). Today's broiler chickens mature in 6 weeks instead of 3 months (Cahaner and Siegel, 1986; Chambers et al., 1981).

The positive gains achieved through applying genetic theory to animal improvement do not necessarily reflect a strategy for coping with different goals that change with time. Instead, the emphasis is

AN EXAMPLE OF BREED UNIQUENESS

North Ronaldsay sheep eat seaweed on Linga Holm, one of the Orkney Islands of Scotland. The island is owned by the Rare Breeds Survival Trust and run as a sanctuary for this breed. Credit: Howard Payton, Cotswold Farm Park.

A much larger variety of livestock breeds exists than is used in commercial agricultural operations. Thousands of years of adaptation to region-specific environmental conditions and more recent domestication have created diverse and unique traits in many breeds, which today are rare or endangered. The potential loss of rare or endangered breeds is of concern to livestock breeders, who may need their special characteristics to adapt commercial breeds to unforeseen needs, and to researchers, who could study them and expand society's understanding of domestication, selection, genetics, and evolution.

One such rare breed of sheep, indigenous to an island in the Orkney archipelago off the northeast coast of Scotland, has adapted to a unique on identifying selection objectives that represent current economic situations in the industry and on attempting to maximize rates of genetic change to achieve those objectives.

This approach can also have a negative effect on maintaining genetic diversity. Superior animals are identified and, through reproductive technologies, can produce large numbers of progeny. Certain breeds are recognized for their superior performance and are more widely used and more rapidly improved. Breeds that are not economically competitive, or that are perceived as being less profitable, are less popular and are not maintained by producers. A 1982 survey in western Europe identified more than 700 distinct breeds of cattle, goats, horses, pigs, and sheep, about 240 of which were classified as endangered (Maijala et al., 1984).

---

set of severe environmental conditions and management practices. It serves as an example of a livestock breed with an exceptional physiology that merits preservation because its genetic characteristics cannot be found elsewhere.

The North Ronaldsay sheep survive exclusively on a diet of seaweed on an island shoreline buffeted by the North Sea's cold winds and winter storms. They have fine bones, a naturally short tail, and a small body, one-third the size of sheep in commercial production. Characteristic of all primitive sheep breeds, the North Ronaldsay displays a wide range of wool color, from white to grey or blue-grey and occasionally brown or black.

The islanders of North Ronaldsay erected a protective, circumscribing seawall 160 years ago to prevent foraging sheep from damaging the iodine-rich seaweed surrounding the island. When the demand for cropland increased the sheep were placed outside the seawall, and seaweed became their primary source of forage. Then, as now, islanders harvest the sheep for meat and wool.

Excluded from all sources of grass forage, the North Ronaldsay sheep developed distinct foraging preference and behavior that set them apart from other breeds. To obtain all nutritional requirements from limited fresh water and abundant kelp beds along the shore, they have mastered the physiological challenge of handling elements, like sodium, present in excess in their environment as well as obtaining important trace elements, like copper.

The copper concentration present in *Laminaria,* the most preferred kelp, is one-third the concentration found in terrestrial herbage. Other breeds found in Scotland, which normally feed on grass or hay, would

(continued)

Animal breeders have recognized that selection objectives change over time, but they have always assumed that sufficient genetic diversity would remain to permit changes in selection objectives. However, as technological advances improve the ability to select for particular genetic traits, the potential of losing related, unselected genes increases. Consequently, the capacity for an intensively selected trait to be further manipulated through selection and improvement may be compromised.

Reproductive Technologies

Reproductive inefficiency, or the inability to produce regularly the optimum number of offspring, is a costly and economically limit-

---

die from lack of copper if fed *Laminaria*. However, the North Ronaldsay sheep are so well adapted to the low concentration of copper present in *Laminaria* that higher copper levels are toxic to them.

The North Ronaldsay sheep are also very salt tolerant. Foraging on seaweed directly from the sea subjects the sheep to high concentrations of sodium and other salts. With no source of fresh water for much of the year, their capacity to withstand saltwater is unique. The physiological or biochemical mechanism by which this adaptation is possible has yet to be studied and understood.

These sheep have also adapted to the challenges of living and feeding on rocky island shores where the sea is a constant threat. While other sheep breeds have a diurnal feeding pattern, the beginning determined by sunrise and the end sunset, the behavior of the North Ronaldsay sheep is linked to the tidal cycle. The sheep begin to feed about 3.5 hours after high tide when the beds of *Laminaria* are exposed. Feeding ends 4 hours later, shortly after low tide, when the tide begins to advance and the likelihood of being stranded at sea increases.

Until recently the entire population of North Ronaldsay sheep, about 3,000 individuals, existed exclusively on one island. The entire breed was at risk. Sudden environmental degradation, like a catastrophic oil spill in the North Sea, could destroy its food source and habitat. Disease could decrease its number, thereby reducing the genetic diversity within the population.

In 1973, a population of 150 North Ronaldsay sheep was established on the island of Linga Holm by the private, then newly formed Rare Breeds Survival Trust of Great Britain. This reserve sheep flock has now grown to 500 individuals. The physical environment of Linga Holm is similar to North Ronaldsay, but grass forage is available from ing problem for animal industries (Pond et al., 1980). Reproduction is influenced by genetics, nutrition, disease, and environmental factors, such as temperature and photoperiod. Each of these factors can be modified to enhance productivity and profit, but improvements can rarely be realized with only one modification. For example, highly prolific breeds of sheep and pigs exist, but their use in commercial production may be contingent on the ability to provide improved prenatal nutrition to the dam, increased attention to the offspring at birth, and supplemental postnatal nutrients to both offspring and dam. Highly productive dairy cows and layer hens usually require large amounts of feed to produce high levels of milk or eggs.

Where the production environment is ideal, reproduction is limited by physiology. The situation is particularly clear in cattle, which

---

the interior of the island. The sheep still prefer *Laminaria* and other types of kelp. However, other groups, brought to farms in England and Scotland, that have less access to kelp may change their forage preference. Over time, the genotype and phenotype of these splinter populations may be expected to diverge from that of the original in response to new environmental conditions and management practices.

Although not of great commercial value, the North Ronaldsay sheep possess other characteristics that may be of interest to local breeders. The ratio of weight to pelvic dimensions in this breed produces fewer problems during birthing compared with other breeds. Introduction of this trait to commercial sheep on the Orkney Islands may decrease ewe mortality. Commercial sheep introduced on these islands often die because of a lack of grass forage during winter storms. North Ronaldsay ewes crossbred with commercial meat rams produce lambs that are able to finish on *Laminaria* to carcasses of commercial value. These characteristics may decrease losses and increase productivity for farmers in the Orkney Islands.

The remarkable breed improvements realized in recent years were possible, in part, because of the enormous genetic variation developed over centuries of natural selection and husbandry practices. Further enhancement of commercial livestock may be necessary as environmental conditions change, market demand varies, or production mechanisms are modified. Future study of salt tolerance and copper metabolism may benefit from examining the rare, physiological model of the North Ronaldsay sheep. Preserving the unique qualities of this sheep and other diverse livestock breeds will ensure a wealth of genetic resources for future use in basic scientific research and the advancement of the agricultural sciences.

duce one calf per cow each year but may calve less ese low reproductive rates have a major limiting effect on efficiency. They especially limit the effectiveness of ction programs by reducing the number of progeny that can be expected from an individual mating. To achieve an acceptable number of births each year, many females must be bred.

Advances in reproductive physiology and cryobiology have generated the capability to produce large numbers of offspring from only a few parents. Techniques that have particular implications for genetic diversity include the capability to collect and cryopreserve germplasm, particularly spermatozoa and embryos, and the ability to manipulate the breeding of parental stocks through artificial insemination or embryo transfer.

Application of these technologies can increase diversity in a population through the introduction of new genes, but it may also narrow diversity where introduced spermatozoa, ova, or embryos overwhelm or replace local breeds. Semen can be collected from a single male, frozen, and later used to produce tens of thousands of progeny. Techniques for collecting, manipulating, and preserving oocytes, ova, and embryos, although more recent, are rapidly expanding. In cattle, up to a hundred offspring can be produced from a single female, and within a few years in vitro fertilization will likely increase the number of possible offspring to thousands per cow. Techniques have also been developed to split individual embryos and to clone embryos by nuclear transplantation. Cryopreservation using liquid nitrogen permits long-term storage of gametes and embryos as well as shipment worldwide.

As these techniques become increasingly cost-effective, they will be applied in more environments and production systems. Their effects on rates of population improvement have been and will increasingly be profound and favorable. Their potential negative impacts on genetic diversity and the potential for future genetic change have not been fully assessed, but they could be substantial and must be recognized in the design of breeding programs.

Changes in Production Systems and Nutrition

Technological and management advances in livestock production are far more widespread in developed countries than in developing countries. Particularly for dairy cattle and poultry, and to a lesser extent for pigs, industrialized systems of livestock production have been established. Stocks with very high genetic potential for specific traits are raised in high-input, carefully managed environments. Ad-

The Red Star People's Commune in the People's Republic of China produces 10,000 ducks per month for the nearby Beijing market. Credit: Food and Agriculture Organization of the United Nations.

vances in the understanding of nutrient requirements to support high levels of production are rapidly implemented through supplemental feeding. Endemic diseases are controlled by regular programs of vaccination and antibiotic administration. Under these controlled conditions, genetic uniformity is an advantage because superior genotypes, under a stable, improved environment, will maximize production.

The development of uniform stocks, however, increases vulnerability to genetic disaster. For example, most of the world relies on the industrial stocks of a few multinational corporations for eggs and broiler meat (Mason, 1984). A particular disease may develop that could decimate these populations. A similar situation exists for commercial turkeys, which also experienced declining reproductive abilities as size and muscling increased. Artificial insemination is now universally used in the breeding of industrial stocks. Concurrently, the fertilizing ability of semen, number of eggs per female, and egg hatchability have declined (Mason, 1984).

The ramifications of this trend are considerable for developing countries and include the replacement of small, near-subsistence farming operations with larger enterprises that have greater access to feed and technology. A village economy based on diverse indigenous genetic types is not necessarily inferior to intensive production using highly uniform genetic types when market channels are poorly developed and availability of the necessary inputs is limited. However, increased production must be achieved in conjunction with adequate marketing and distribution systems to ensure access to animal products by all consumers.

Management technologies and exotic livestock have been imported into many developing areas in response to demands for higher production levels. Entire production systems for chickens, including

---

EXTENSIVE AND INTENSIVE PRODUCTION SYSTEMS

Livestock are generally produced and managed in two distinct ways: extensive production, practiced for many thousands of years, and intensive production, in use for only the past few decades. Each system has its place in world agricultural production and must be evaluated in terms of its appropriateness for a given region's social, financial, and natural resources.

Extensive systems are employed on small farms in subsistence production and on large farms that produce considerable quantities of animal products for market. Small extensive systems are primarily found in developing countries, while large extensive farms operating under range conditions may be present in both developed and developing countries. Herd or flock size in small extensive systems is usually significantly smaller than that of intensive or large extensive systems, which can accommodate thousands of animals in a year. Intensive operations are found predominately in developed countries on specialized, one- or two-product farms.

Extensive and intensive production systems differ in many other respects as well. Typically, in small extensive production systems, animals obtain forage and water on open pasture and are adapted to survive on available nutrients and trace elements. Livestock in large extensive operations may receive supplemental feeds, such as hay, in addition to range forage. Animals in intensive production systems are given water, fortified mixtures of feeds to enhance desired production characteristics, like increased carcass weight, and supplemental injections of vitamins and trace elements as needed.

Housing is also addressed in varying ways. Animals in extensive operations are more exposed to the elements. Housing in small extensive systems, if provided, is of natural materials, such as straw bedding

(continued on page 36)

management technology, genetic stocks, and sometimes feed supplies, have been successfully transferred. Production systems for pigs, including high-producing genetic lines, are also being transferred. For these species, the trend is to replace indigenous stocks and production methods completely. Purebred dairy cattle have been exported from temperate to tropical regions with varying degrees of success in drier areas but usually with disastrous results in the humid zone. The crossbreeding of indigenous stocks with imported ones, up to certain levels of exotic inheritance, have usually been successful (Cunningham and Syrstad, 1987). However, this result largely de-

This commercial feedlot contains beef cattle that can be fed high-energy feeds to increase their weight rapidly, an example of intensive production. In contrast, grass-fed cattle kept in pastures until they reach a desirable slaughter weight is an example of extensive production. Credit: Gordon W. Gahan, ©1992 National Geographic Society.

pends on the availability of adequate nutrition to support greater milk production and a structured breeding program to stabilize the optimum gene mix.

The demand for greater rates of production and aggressive marketing practices on the part of exporters can lead to the introduction of unsuitable animals without scientific and technical information about their adaptability for the region. Management technologies (for example, climate control) and physical inputs (for example, high-quality feeds) are necessary to maintain productive exotic stocks. The long-term efficiency and sustainability, as well as the social and eco-

---

for ground cover and branches for makeshift corrals, collected from the immediate surroundings. Large extensive systems may provide wooden or concrete shelters or paddocks. In intensive systems, high-capital investments are required for housing construction and equipment, which may include pens or cages in enclosed buildings with ventilators, thermostats to control temperature, heated floors, insulated roofs, and artificial lighting.

Animals in extensive production systems adapt to regional climate, temperature extremes, and seasonal variations in photoperiod, while animals kept in artificial, intensive production settings are kept in a managed or artificial environment. Some intensive production environments are deliberately manipulated to increase production. Layer chickens, for example, may be placed in a 26-hour light-dark cycle to optimize egg production.

Little or no effort is made in extensive production systems to provide preventative care or symptomatic treatment of disease or parasitic infection. The success of intensive production systems, however, depends on the provision of preventative inoculations and prompt veterinary treatment of sick or diseased animals. The concentration of animals in intensive systems requires rigorous sanitary measures to prevent the spread of disease. For example, wastes are removed regularly and animal living areas cleaned. The udders of dairy cattle in intensive systems are also given an antibacterial rinse after each milking to prevent the spread of infections of contagious pathogens.

The genetic composition of the animals used in the production systems is markedly different. Small extensive production systems may have relatively homogeneous (and often inbred) individuals within each production unit, but the number of production units (herds or flocks) may be large. Animals in different production units may represent nomic implications, of imported production systems must be closely examined. Improvements in livestock production can be explored and implemented as appropriate for a region, but until their relative values and costs are thoroughly understood, endangered indigenous stocks should be protected.

Currently the survival of many indigenous populations is threatened by the introduction of animals from temperate areas. Indigenous stocks are crossed with or, more important, replaced by imported animals. Under conditions of controlled and artificially enhanced reproduction, imported breeds can replace indigenous stocks at a rapid

---

various indigenous breeds. Genetic differences among production units and periodic exchange of breeding animals maintain genetic diversity within the entire production system. Animals in intensive systems, by contrast, are genetically similar individuals from one or two commercial, highly productive breeds that are selectively mated according to a farmer's breeding program. Large extensive operations may have several breeds or only one or two.

Exposure to disease, parasites and regional extremes of temperature, humidity, and photoperiod has produced livestock capable of surviving and efficiently producing food and fiber from available resources, relatively unaided by humans. Animals in intensive production environments are shielded from the extremes of the natural environment and protected from most diseases. If the artificial production environment were not maintained, animal productivity would decrease. A new disease strain is likely to cause a higher infection and mortality rates in intensive production systems because animals live in closer proximity, in greater numbers, and have less genetic diversity.

Advantages and disadvantages exist in both types of production systems. Extensive systems have the advantage of requiring less labor and capital investments. Native breeds used in extensive systems have lower rates of production than commercial breeds, but they are less vulnerable to disease and do not depend on costly production inputs and artificial environments to maintain production. Intensive systems generate greater quantities of food and fiber than extensive systems by maximizing the production capabilities of a few, highly productive animal breeds. Extensive systems represent an affordable way for farmers to produce food and fiber efficiently in areas where the high capital costs and input demands of intensive management practices would be prohibitive.

rate. Generally, the relative merits of indigenous and imported stocks are only superficially weighed when crossbreeding or upgrading programs are being considered. Comparisons of performance should be made under the conditions in which the animals will be maintained. In many cases, a mixture of imported and indigenous genes will be required to produce suitable crossbreeds. However, indigenous stocks should be protected until the best crossbreeds can be determined. Until then, breed replacement runs the risk of losing genes for later adaptation.

RATIONALES FOR CONSERVING LIVESTOCK GENETIC RESOURCES

To some extent, the loss of less competitive livestock can be considered a normal part of the livestock improvement process. Natural selection favors animals that are best suited to a particular environment. The genetic composition of a species is thereby adjusted to new requirements over time. Artificial selection simply redirects and accelerates this process for the presumed benefit of society. Most dependent on genetic diversity are the ruminants, because they must adapt to a wide range of climatic conditions and nutrition levels, and because they must interact closely with their environment to forage for feed. They have acquired great diversity that could be reduced through the misapplication of new technologies for propagating and disseminating germplasm.

A variety of unpredictable factors could change current demand for animal products and in turn alter animal production systems. For example, many of today's consumers demand leaner meat and less fat in their diets. Pigs that are high in fat, once prized for lard production, have been replaced by modern breeds and hybrids that meet current market and price structures based on lean yield (Alderson, 1990a).

The chief reasons for conserving livestock genetic resources have economic, scientific, and cultural and historical bases. These factors are discussed below.

Economic Reasons for Conservation

The strongest basis for conservation is inherently practical: preserved germplasm may contribute to future increases in the efficiency of livestock production. Production demands and conditions have changed in the past 50 years, and the ability of livestock producers to respond has been due in part to the ready availability of a wide

The Pantenero swamp cattle of the Pantenal in Brazil are adapted to the extreme heat and humidity of the region. Credit: Elizabeth Henson, Cotswold Farm Park.

range of genetic variation. In 1940, for example, 25 million cows in the United States produced 109.4 billion pounds of milk. In 1990, 10 million cows produced 148.3 billion pounds of milk (P. Vitaliano, National Milk Producers Federation, Arlington, Virginia, personal communication, June 1991). Greater production efficiency was achieved mainly through improved feeds and the selection of breeding stock with higher capacity to produce fluid milk.

The economic benefits of preservation relative to its costs are difficult to assess without knowing the nature and degree of future change, which is not easily predicted. The development of pathogen strains that are resistant to specific antibiotics, for example, could require access to germplasm not available within a dominant breed. Decisions about what to preserve are difficult because the costs of maintaining breeding herds or populations are high.

Smith (1984a), however, demonstrated that preservation programs based on cryopreservation have very low costs relative to the value of the livestock industry they are designed to protect. He estimated that the yearly costs for satisfactorily maintaining a breeding stock range from £3,000 ($4,000) for sheep or chickens to £12,000 ($16,000) for pigs. Although the costs to collect and place semen or embryos in cryogenic storage may also be high, they are incurred only once and yearly storage costs are low (Smith, 1984a,b). Smith (1984a) thus concludes that the use of cryopreservation is sound because of the large potential returns relative to the low continuing costs of maintenance.

Insights into the molecular makeup of the genome advanced quickly in the 1980s. The foundation is being laid for mapping the genomes of all livestock species and for gaining a better understanding of gene action. Although an understanding of variation at the genomic level is only beginning to develop, molecular techniques offer tremendous promise for identifying and manipulating valuable genes for a marketable trait or a physiological function. Biotechnological advances may allow identification of individual genes that can have positive effects on production, adaptation, and disease resistance traits. The ability to find useful genes in otherwise undesirable stocks and to transfer these genes mechanically to improved stocks will augment genetic diversity. This potential provides a powerful argument for maintaining currently uncompetitive stocks.

Methodologies for studying the genome at the molecular level are not, as yet, being routinely applied to livestock populations, even in developed countries. Nevertheless, rapid progress is being made in techniques for gene mapping and for identifying associations between specified segments of the DNA (deoxyribonucleic acid) and primary production traits.

Application of these techniques to animal species will aid in discerning and characterizing diversity among various populations, but it may have both positive and negative effects on maintaining diversity. Marker-assisted selection involves the use of molecular genetic information to supplement traditional performance information and, in some cases, can increase rates of genetic improvement. Unfortunately, marker-assisted selection also has the potential to increase genetic uniformity and could produce an additional decline in genetic diversity in highly selected populations.

Documentation of endangered breeds in some developed countries has been done by nongovernmental private interest groups, such as the Rare Breeds Survival Trust (RBST) in England and the American Minor Breeds Conservancy (AMBC) in the United States (Alderson, 1990a). In other developed countries, such as France as well as the nations of Eastern Europe and the Commonwealth of Independent States, government programs have been established (J. Hodges, consultant, Mittersill, Austria, personal communication, March 1992). Generally, endangered breeds have lower profitability under prevailing production and marketing conditions; therefore, substantial investment in their preservation by commercial interests is unlikely.

Public support to preserve the unique and most endangered of these breeds may be required. Some breeds in developing countries have become extinct, but the extent of risk and the rate of decline to other indigenous breeds are known in only a few developing countries (J. Hodges, consultant, Mittersill, Austria, personal communication, March 1992). Economic arguments for preservation are more readily justified for indigenous livestock that are not well characterized because of specific genes they may carry or their potential genetic contributions.

Scientific Rationale

Diversity is the base on which genetic research depends. Preserved genetic diversity can be of value to many scientific investigations. Researchers may use genetic variants to clarify mechanisms of development and physiological controls, patterns of domestication and migration evolution and speciation, and other biological questions. For example, the annual number of lambs per ewe vary greatly among sheep breeds. Certain sheep breeds from countries such as Australia, Bangladesh, China, Finland, Indonesia, Ireland, Poland, and the United Kingdom have high lambing rates. Through crossbreeding, they provide the means for genetically transforming ewe prolificacy. In the Australian Booroola Merino, a single major gene produces an increase in ovulation rate, but its operation is not completely understood and requires further study (Bindon and Piper, 1986; Thomas, 1991).

Special genetic stocks of livestock species are also essential research tools. For example, inbred lines of chickens developed at the University of California at Davis are used worldwide for research on immunology and disease resistance of chickens (Office of Technology Assessment, 1987). Other beneficial uses of preserved germplasm will emerge as genome maps are created and as scientists learn to identify and manipulate specific genes involved in the mechanisms of growth, reproduction, and disease.

Cultural and Historical Rationale

Preservation programs can provide visual evidence of a nation's heritage. Just as historical qualities are valued in buildings and geographic sites, the same regard can be extended to animal populations that are representative of antique or heirloom breeds or populations. In restorations of historic villages, for example, old (and often rare) breeds are desired for authenticity. Farm parks, in which a large number of endangered and unusual breeds are maintained for the public, have proved popular in England, where 12 farm parks are now operating in association with the RBST. Similar programs have begun in other countries (Alderson, 1990a). Furthermore, some animals are used for recreation and in animal shows, exhibits, and fairs.

One of the poultry stocks identified by the AMBC as being endangered is Dominique chickens, perhaps the oldest of the chicken breeds in the United States and present as a barnyard chicken since colonial days. They have black and white feathers and a rose comb and are featured in American folk art, music, and literature. The few surviving today in hobbyist exhibition flocks are highly inbred and not very vigorous.

Some livestock breeds characterize a way of life and provide specific products in areas around the world. For example, six native cattle breeds of southern Spain are in serious danger of extinction—the fighting bull was derived from these cattle. Among the endangered breeds are the dual-purpose Berrenda breeds, used for meat production and as steers to assist in the handling of fighting bulls (Rodero et al., 1990).

Although the community value of a certain stock for its cultural heritage role is difficult to assess, its economic value is evident from the financial support provided by visitors to farm parks. Thus, livestock can, arguably, be accorded the same consideration as other endangered reminders of national heritage.

APPROACHES TO LIVESTOCK CONSERVATION

Numerous reports agree on the need to develop national and international efforts to preserve and manage livestock genetic resources (Alderson, 1990a; Council for Agricultural Science and Technology, 1984; Food and Agriculture Organization, 1984a,b; Hodges, 1987; Office of Technology Assessment, 1985, 1987; Smith, 1984b; Wiener, 1990). However, the approaches to conservation differ widely within the livestock community. They hinge on different perspectives about the importance of traditional livestock breeds, including indigenous breeds, to long-term genetic improvement programs. These perceptions also vary between developed and developing countries.

In general terms, these approaches can be divided between two broad categories: (1) utilizationist, the primary aim being immediate use of available genetic resources, and (2) preservationist, the primary goal being long-term preservation of genetic resources for unknown future use. Proponents of both approaches recognize that conservation of genetic diversity is economically justified only to the

Oxen are still used today for plowing rice paddies in Bangladesh. Credit: Steve Raymer, ©1992 National Geographic Society.

extent that the preserved germplasm can reasonably be expected to make a contribution to future livestock improvement programs. Both also recognize that preserving breeds that are not currently economically viable has a place in livestock improvement systems. Differences between the two approaches largely pertain to perspective and emphasis, but they are real and strongly influence the practices favored for managing genetic resources. The use and definition of the terms *utilizationist* and *preservationist* are not perfect or exclusive. Rather, they serve to illustrate the diversity of approaches to genetic conservation of livestock.

The Utilizationist's View

The utilizationist gives the highest priority to using available genetic resources to improve current livestock populations. Increases in the rate and efficiency of livestock food and fiber production are paramount. Existing populations are crossbred in carefully planned programs to achieve new and potentially more valuable gene combinations. The loss of breeds as distinct identities is not generally viewed as a matter of concern, as long as the genes that make those breeds potentially useful are retained in commercial stocks.

This approach also recognizes the fallibility of planners, producers, and scientists. Endangered stocks must be preserved during the period of evaluation, crossing, and progeny assessment, and adequate genetic diversity must be maintained within commercial stocks to allow responsiveness to future changes in selection goals. Stocks with documentable, unique biological characteristics must also be preserved for future use. However, the utilizationists place no particular priority on preserving stocks that have not contributed to improving current populations or that have no documentable, unique biological characteristics that they foresee as having potential future importance.

In a report on animal genetic resources conservation in the United States, the Council for Agricultural Science and Technology clearly presents the utilizationist's view. It states that two kinds of animal germplasm merit priority for preservation (Council for Agricultural Science and Technology, 1984). They are the following:

- Breeds threatened with deterioration or extinction that have either known superiority for one or more traits or genetic uniqueness that could be important in meeting unforeseen needs.
- Stocks that are useful for research or potential commercial exploitation because of unique genetic traits.

The report further states, "priority in germplasm preservation should favor stocks of known superiority in specific traits over those that are merely different from genetic material that is widely accepted" (1984:29). It indicates that special efforts to preserve rare livestock and poultry breeds may not be justified if they do not have identified, potentially useful biological characters.

The Preservationist's View

The preservationist regards existing breeds as unique genetic entities representing genes and gene combinations that have evolved over considerable periods of time to fit unique environmental and production conditions. This approach emphasizes the value of preserving the widest possible spectrum of genetic diversity, because it is not possible to predict future needs. Breed preservation is regarded as the most practical method of gene conservation. The crossbreeding of endangered breeds with more popular stocks is considered to be a threat to the survival of endangered stocks. It is acceptable only when the identity of the original stock as a purebred line has been preserved. The cultural and historical importance of breeds is also given high priority.

The preservationist's view recognizes the potential value of breeds to livestock development, but it places higher priority on preserving existing breeds. Future use of genetic resources in crossbreeding and selection programs at the commercial level remains the primary justification for conservation, but the inability to predict future needs and, therefore, to choose which breeds are most worthy of preservation remains central to the rationale for this perspective.

The Cotswold Farm Park in England gives three reasons for saving rare breeds that do not appear to meet today's production standards. First, they are part of society's heritage. Second, their characteristics could make them ideal research subjects. Third, and most important, they provide livestock breeders with a broad pool of genetic diversity to meet unforeseen needs (Cotswold Farm Park, 1985).

The Committee's View

The utilizationist emphasizes the characterization of existing breeds and the evaluation of the most promising as components (either purebreds or crosses) of current production systems. Only the most promising breeds or those that represent documentable biological extremes should be preserved. Improvement of existing populations should meet current and perceived future needs. The preservationist's view, by contrast, calls for preserving the greatest possible number of breeds as purebreds. Crossing to form new populations should be accompanied by the protection of breed identity. The preservationist emphasizes the inability to predict future needs and, therefore, to create a priority list of candidate breeds for preservation.

Reconciliation of these two views must begin with the recognition of their common goal: sustained long-term improvement of livestock for use by an expanding global population. The main conflict is in how each view establishes priorities for conservation and how each assesses future livestock needs. Current human needs for food and fiber are pressing and will continue to increase. These needs must be met. The goal of this report is not to validate one position at the expense of the other, but to present a position that is acceptable to these divergent views for building coordinated efforts toward the common goal of conserving global livestock diversity.

The committee agreed that one goal of genetic resource management programs must be to maximize rates of genetic improvement in commercial stocks. It also endorsed the view that some form of genetic resource conservation is an essential component of long-term livestock improvement programs. Differences of opinion existed within the committee on the amount of resources that should be devoted to breed preservation. It was agreed that preservation should, at a minimum, address breeds that are of (1) potential economic value, or are (2) endangered and represent types with documentable, unique biological characteristics. For a national program, this approach would encourage maintenance of adapted indigenous types as pure lines while they are being compared with imported stocks and while their optimum role in breeding programs is being determined. It would also encourage long-term preservation of genetic extremes that could provide material valuable in future commercial stocks.

The establishment of priorities and the determination of which breeds or populations are to be preserved will depend on several considerations, including the resources available, national and international needs, and the genetic similarity among available breeds and populations. Development of an understanding of population structures and genetic relationships among breeds or strains can aid the setting of priorities.

The committee thus concluded that responsible management of livestock genetic resources involves aggressive programs of livestock improvement as well as strategic programs for conservation.

RECOMMENDATIONS

The development of comprehensive national and global strategies for preserving and using animal genetic resources will require the integration of livestock improvement programs and programs for conserving livestock genetic resources. To date, improvement programs have taken precedence over conservation programs, largely because of the quicker and greater return to investment. Yet, a long-term approach to increasing global food resources demands that conservation become an integral part of programs for animal improvement.

Mechanisms must be put in place to ensure that genetic diversity of the major livestock species is maintained to support improvements in production efficiency and to accommodate future changes in selection goals.

Global human populations will continue to increase in the foreseeable future. Global capacity for food production must, therefore, also increase. Most major livestock populations of developed countries have been subject to intensive artificial selection for a period of less than 50 years. Scientists cannot predict what level of genetic diversity will be required to support genetic manipulations in livestock populations over periods of many hundreds of years. Yet, action is required now to maintain genetic diversity, and it must be undertaken responsibly, regardless of current limitations in understanding the basis of genetic diversity. These mechanisms for action could take many forms, depending on national and international needs and resources, but their common objective would be to conserve the livestock genetic resources essential to agriculture and society.

Clear national and international policies are needed to prevent further losses of potentially important animal genetic resources.

Agricultural research policies have focused exclusively on livestock improvement programs. They must recognize that conservation is a vital part of these efforts. In developing countries, the responsible handling of indigenous breeds must be given a high priority.

Priority for preservation should be given to species, breeds, or populations that are both at the greatest risk of loss and that appear to have potential for future use in livestock improvement programs.

In most countries the resources available to preserve animal genetic resources will be limited. In addition, not all animal species or particular breeds within a species will have a significant economic role in other countries. Preservation efforts must focus first on those species and breeds that are at risk and that have potential for contributing to future economic or social development in the country. Although it may be desirable to save all endangered breeds, limited resources will necessitate the setting of priorities, based on perceived future economic contribution and biological uniqueness.

2

Essential Conservation Considerations

The collection, transfer, and use of animal germplasm are influenced by several fundamental considerations. In contrast to plant germplasm collections, little or no collecting for breeding purposes is done from free-ranging or wild animal populations. Almost universally, individual domestic animals or their gametes or embryos must be purchased. The maintenance of adequate numbers in representative live populations will generally not be economically feasible for livestock species kept solely for preservation. When cryopreservation is used, semen or embryos must be processed immediately after collection to ensure their survival. Regeneration of cryopreserved germplasm requires an established protocol in which host females are available, and in the proper physiological condition, for embryo transfer or artificial insemination. Internationally, the movement of animal germplasm may be limited by strict quarantine regulations.

These constraints must be addressed in planning and instituting conservation programs. This chapter examines the technical elements of managing the genetic resources of livestock including preservation criteria, sampling strategies, and preservation methods.

CRITERIA FOR CONSERVING POPULATIONS

An animal germplasm conservation program will require decisions on the populations, which can be a particular breed or stock, to be preserved and the methods to be used. Factors that will influence those decisions include the status of a particular population and its uniqueness or economic importance, both current and potential.

Status and Vulnerability of the Population

To evaluate accurately a population's status, it is necessary to have a census of its total number and an understanding of its dynamics, that is, whether the population size is stable or changing and the rate of any possible change. Rough estimates of the size at which a population in a developed country should be considered endangered have been provided by Maijala et al. (1984) and are summarized in Table 2-1. A more detailed scheme for assessing populations and determining appropriate conservation or management actions has recently been published by the Food and Agriculture Organization (FAO) of the United Nations (Henson, 1992); it is depicted in Figure 2-1.

In the United States the American Minor Breeds Conservancy (AMBC) uses the categories rare, minor, watch, and feral to classify the status of a livestock breed that may become endangered based on the numbers of individual registrations per year (Henson, 1985). (See Table 2-2.) These categories are defined as follows:

- Rare—cattle and horse breeds with less than 200 registrations per year; sheep, goat, and pig breeds with less than 500 registrations per year.
- Minor—cattle, sheep, goat, and horse breeds with less than 1,000 registrations per year. Pig breeds with less than 2,000 registrations per year.

TABLE 2-1 Recommended Sizes at Which Livestock Populations in Europe Should be Considered Endangered

Livestock Species	Size of Population	Direction of Change	Number of Breeding Females	Number of Breeding Males
Cattle	1,000–5,000	Declining	<1,000	<20
Sheep and goats	500–1,000	Declining	<500	<20
Pigs	200–500	Declining	<200	<20

SOURCE: Maijala, K., A. V. Cherekaev, J.-M. Devillard, Z. Reklewski, G. Rognoni, D. L. Simon, and D. E. Steane. 1984. Conservation of animal genetic resources in Europe. Final report of an EAAP (European Association of Animal Production) working party. Livestock Prod. Sci. 11:3-22. Reprinted with permission, ©1984 by Elsevier Science Publishers.

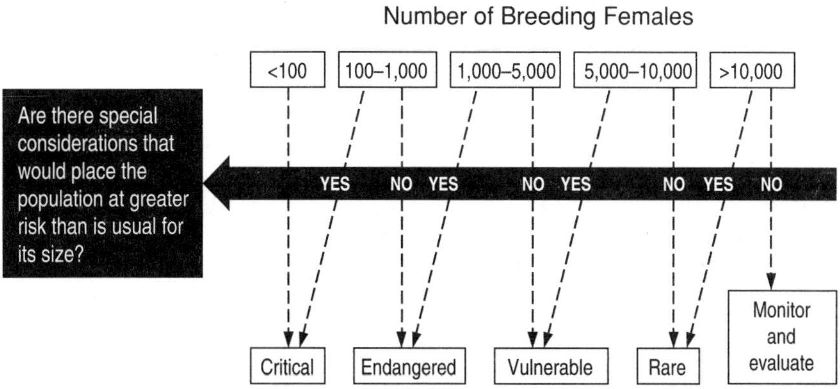

FIGURE 2-1 Proposed scheme for classifying livestock breeds and populations identified as in need of preservation. Note: Greater detail and protocols for determining specific actions can be found in Henson, E. L. 1992. In Situ Conservation of Livestock and Poultry. FAO Animal Production and Health Paper No. 99. Rome, Italy: Food and Agriculture Organization of the United Nations.

- Watch—breeds whose registrations over a 25-year period have shown a steady decline or where registrations are less than 5,000 per year.
- Feral—stocks known to have been running wild for at least 100 years with no known introductions of outside blood.

The information contained in Tables 2-1 and 2-2 is based on the numbers of breeds in developed countries and reflects subjective judg-

TABLE 2-2 Categories Used to Classify the Status of Livestock Breeds by the American Minor Breeds Conservancy

	Maximum Number of Breed Registrations Per Year				
Status	Cattle	Horses	Sheep	Goats	Pigs
Rare	200	200	500	500	500
Minor	1,000	1,000	1,000	1,000	2,000
Watch[a]	5,000	5,000	5,000	5,000	5,000

[a]A breed may also be placed in this category if a steady decline in registrations has occurred over the preceding 25 years.

SOURCE: Henson, E. L. 1985. North American Livestock Census. (Updated May 1988.) Pittsboro, N.C.: American Minor Breeds Conservancy. Reprinted with permission, ©1985 by American Minor Breeds Conservancy.

ments about the necessity for concern. For developing countries, other factors may affect the risk of survival and must be considered when evaluating risk. For example, a population may be at serious risk, even if it numbers several thousand, because of a drought or disease epidemic. Geographic isolation or fragmentation may also affect risk. Extensive crossbreeding, particularly if artificial insemination is used, can rapidly change the genetic composition of a given population. Unique populations can be genetically diluted through such activities. The FAO recommends that populations in developing countries be evaluated as candidates for preservation when the number of breeding females drops to about 5,000 within a total population size of about 10,000 (Hodges, 1990b).

Uniqueness and Importance of the Population

Populations that have developed in isolation may possess important genetic traits not found elsewhere. Many indigenous breeds are found only in the region in which they originated. They should receive special consideration for inclusion in a national conservation program, particularly if the environmental conditions encountered in the region where they were developed are unusual. Alternatively, if

THE SIGNIFICANCE OF EFFECTIVE POPULATION SIZE

The size of a population is simply the number of individuals in it. However, scientists are more concerned with the flow of genes within the number of individuals contributing gametes to the next generation—the "effective" parents in the population.

The relative number of effective parents of each sex in a population is important to assess effective population size (N_e). If, for example, there are few breeding males in a population, then the effective size will be much smaller than its actual population size. In general, anything that affects the breeding capacity of individuals in a population or the survival potential of their offspring can alter effective population size. Thus, selection, whether natural or part of a breeding program, can be important.

The importance of N_e to livestock is seen in a hypothetical example of world populations of black and white cattle—the familiar Holstein (American) and Friesian (Europe) dairy cattle (Goddard, 1991). While providing significant benefits in terms of wide access to the best producing animals, international trade of the germplasm of these animals through the exchange of cryopreserved semen and embryos has the a breed is also found in neighboring countries, or if several breeds appear equally well adapted to some environmental niche, the need for protective measures may be lower.

Measures of genetic distance may help identify candidates for preservation. For example, the close relationship of an endangered breed to another nonendangered breed could lessen the urgency to preserve it over another, more distantly related breed.

STRATEGIES FOR SAMPLING POPULATIONS

Once a population has been identified for preservation, the next step is to determine how to sample it. Sampling strategies must take various factors into consideration, including geographic distribution of the population and subpopulation structure, if known. Sampling biases can be minimized by using a stratified system of sampling, identifying populations (breeds) and subpopulations (herds or stocks), and taking samples (of unrelated individuals and at random) in a systematic manner. The preserved samples should reflect the proportional balance of subpopulations present. With stocks that are rare or endangered, there may be little choice for sampling, and essentially all individuals available may be included. (Factors that af-

---

potential to greatly reduce their effective population size. Goddard notes that the potential exists to manage these cattle as a single global breed. Under this scenario, the N_e for the global population of black and white cattle could be as low as 80, even though the estimated world population would include several million milking cows. A global program would likely use sperm from very few bulls and, through embryo transfer, limit the number of females contributing to future generations.

The global breed scenario may be an extreme example, but it illustrates the potential impact of widely distributing the genes of a limited number of individual animals. Attainment of small global N_e will be moderated by even slight variation in breeding objectives or environment because breeders in different countries would then need to use somewhat different breeding animals to meet their goals. Likewise, some would argue that an N_e of 80 may be sufficient to allow reasonable long-term selection responses. Yet long-term propagation of a population at such modest N_e will almost certainly result in some losses in genetic variability. The future importance of such diversity cannot now be quantified, but programs to maintain a reserve of genetic diversity seem warranted in populations being maintained at low N_e.

fect the collection and storage of embryos, including estimates of the numbers required, have been presented by Springmann et al. [1987].)

The number of individuals to be sampled need not be large to prevent loss of genetic variation. For example, the committee supports the conclusions of the Council for Agricultural Science and Technology (1984) and of Smith (1984b) that, if animals are properly sampled, semen from 25 males or embryos or live progeny of 25 males and 50 females would reduce the potential loss in genetic variation in the initial sample to less than 1 percent. With frozen stores, no loss of variation occurs until the store is used, whereas with living populations of small sizes, a continual decline of genetic variation can occur through inbreeding or genetic drift. For rare stocks, all males and females should be sampled if possible.

METHODS FOR PRESERVING LIVESTOCK GERMPLASM

Three primary methods have been used by public and private programs to preserve livestock germplasm: maintaining live populations, cryopreserving gametes and embryos, and establishing DNA (deoxyribonucleic acid) stores. In some cases, keeping breeding populations may be the most practical method. Cryogenic techniques will continue to be improved and developed and, thus, should be considered the best strategy for long-term preservation. Cryopreservation technology can be applied to most domestic species. However, important exceptions are semen from swine and embryos from swine and avian species. Embryos from pigs have been cryopreserved at expanded blastocyst and early hatched stages and pig semen can also be frozen.

DNA is being stored in gene libraries and as an addition to cryogenic stores. Although long-term storage of DNA is likely to be more important in the future, it is not yet a viable method for preserving and using livestock germplasm.

Managing Live Populations

For many countries the most practical way to preserve animal germplasm is to maintain live breeding animals at as many locations as feasible. The herds and flocks maintained by private owners often suffice, but if a given breed becomes endangered, publicly supported efforts may be necessary to develop breeding programs that will control inbreeding and ensure breed purity.

The disadvantages of preserving small living populations include costly facilities and supervision and exposure to several hazards. These hazards include loss due to disease, the increased potential for inbreeding that may result in a greater frequency of deleterious traits, contamination from other stocks, and changes due to natural selection (Smith, 1984b).

As an alternative to individual breed or population preservation, animals representing several different breeds could be maintained in a single interbreeding population. Although this method is suitable for preserving the constituent genes of the various breeds, particularly if many breeds merit saving but financial or other restrictions exist, it has significant drawbacks. The population would serve to maintain single alleles, particularly if they confer a selective advantage, but the gene combinations associated with specific breeds would be broken up and lost. Isolation of the individual gene combinations that were characteristic of a breed would be difficult after they have been combined in a single interbreeding population. Also, special mating programs are needed to avoid loss of some genes through negative selection. Thus it is preferable to preserve gametes and embryos from purebred lines rather than from a pooled population.

Public and Private Initiatives to Manage Live Populations

The conservation of animal genetic resources can involve both public and private initiatives (Crawford, 1990a; Setshwaelo, 1990; da Silva, 1990; Wezyk, 1990). Many governments have established in situ herds in public parks set aside for wildlife conservation. Depletion of the Longhorn breed of cattle in the United States, for example, was prevented by keeping semiferal herds in state parks in Oklahoma. In 1932, the South African government collected and preserved the Nguni cattle in the Natal Parks (Setshwaelo, 1990). Similar approaches are used in France with Rove goats and Camargue cattle and horses, among others. At the Hortobagy National Park in eastern Hungary, government resources and initiatives ensure preservation of Hungarian Grey Steppe cattle, Mangalica pigs, Racka sheep, and poultry (Alderson, 1990a; Henson, 1990; Wezyk, 1990). Many individuals and private groups are interested in maintaining and propagating unique types of animals (Alderson, 1990a; Office of Technology Assessment, 1985). For example, private farm parks, such as those supported by the Rare Breeds Survival Trust (RBST) in the United Kingdom, have proved successful (Alderson, 1990b; Henson, 1990). These working farms keep rare breeds and are financially supported by the public, who pay to enter.

Public subsidies have been used to encourage private on-farm maintenance of rare breeds. In some cases government subsidies

Gloucester Old Spot pigs are an endangered native livestock breed in Great Britain. They are hardy and could be crossed with other breeds to produce commercial pigs that can be kept outdoors. Credit: Cotswold Farm Park.

have been provided to farmers to maintain animals of a specific breed. For example, the regional government of the Province of Quebec, Canada, pays farmers to rear purebred Canadienne cattle (Henson, 1990). The methods of payment and the amount of the subsidy may be

- Per head. The system of paying a per head rate on each animal raised is simple, but it can lead to problems of overgrazing and mismanagement of flocks or herds.
- Production linked. These subsidies seem to be the best controlled and most effective. The subsidy is determined by evaluating the production of the rare breed under normal management and comparing it with the potential production of the replacement breed under the same management system. The farmer is paid a subsidy

equivalent to the difference. With production-linked subsidies it is still essential that the farmer continue to manage the animals well to realize the maximum value from them.

- Male only. This is probably the simplest form of subsidy. Farmers are paid for rearing males and making them available for semen collection and for use in contract matings. It has been used successfully in Britain for maintaining rare strains within cattle and pig breeds, and it demonstrates the feasibility of close cooperation among individual farmers, breed societies, private conservation groups, and farmer-funded organizations.

Cryopreservation

Cryopreservation of semen and embryos complements the preservation of live populations, providing a safeguard when breed numbers are dangerously low or when breeds or lines are likely to be replaced or lost. It does not allow for continued adaptation of the population to changes in environmental conditions, but it does offer the potential to store large numbers of genotypes for indefinite periods of time.

Generally, the costs of collecting and processing samples for cryopreservation are not excessive for domestic animals, particularly for semen (Smith, 1984a). However, they may vary considerably, depending on the particular collection conditions. Once the material is collected and processed, the cost of maintaining it in cold storage is much lower than the cost for maintaining live animals. Furthermore, cryopreserved samples kept in liquid nitrogen can be moved long distances with relative ease.

The successful cryopreservation and banking of animal germplasm requires special expertise. First, the behavior and reproductive physiology of the animal must be known and suitable for control and manipulation, and the cryopreservation procedures must be adapted to the gametes or embryos of the particular species being preserved. Second, the breed must be sampled with care to ensure preservation at the desired level of diversity. Third, females must be available to serve as recipients of cryopreserved germplasm. Finally, secure, long-term storage facilities with trained personnel must be established. Ideally, duplicate samples should be kept at other locations.

Finally, the health status of the donor animals must also be checked and records maintained. Appendixes B and C discuss the risks of disease transmission that can be associated with the collection, storage, and transfer of embryos and semen, respectively. New techniques for washing embryos (see Appendix B) will greatly advance capabilities to store germplasm that is free of pathogens. Before cryopreserved samples are regenerated, the health of the recipients should be determined to ensure that healthy germplasm is not contaminated.

Cryogenic technology may not be routinely applicable to all breeds or in all regions. Embryo collection involves maintaining donor females at suitable facilities and having adequately trained personnel. Hormonal profiles vary among breeds so that hormonal treatments for embryo transfer technology must be developed for individual breeds. Factors such as climatic or dietary stress may affect hormonal response. In Britain, for example, a failure to collect cattle embryos of one breed was apparently due to differences in the physiological responses of the rare breed cows to hormonal superovulation treatment (E. Henson, Cotswold Farm Park, United Kingdom, personal communication, July 1987).

Finally, some animals produce gametes that have higher viability rates than others after freezing. For example, much success has been achieved with cattle semen. In contrast, the cryopreservation of pig semen is feasible but conception rates are poor, resulting in higher regeneration costs. These individual differences may lead to unintended selection pressures on the population.

Genomic Libraries

In some cases the rationale for preserving a particular breed of animals is solely to preserve one or a few traits unique to it. Because variation within species is molecularly based, the phenotype may originate from a single gene or a few genes. It may be possible and advantageous to preserve the desired genetic traits by preserving genes in addition to maintaining live or cryogenically stored semen or embryos. In the future, it is expected that techniques will exist for preserved genes to be transferred into embryos of the same species or into other species, thus permitting the desired traits to eventually be returned to a living animal population. For these reasons, genomic libraries may have an important place as a supplement to frozen semen and embryo banks in a long-term strategy for preserving animal genetic resources.

The value of genomic libraries will be continually enhanced as more is learned about the molecular physiology of animal species. For example, in cattle the gene products of the major histocompatibility complex, a region of the genome that is involved in immune responsiveness, are being characterized (Benyo et al., 1991; Maguire et al., 1990). Differences in the gene products are being noted among breeds and are being associated with differences in production traits. With further understanding of these products, it may be feasible and desirable to transfer genes among groups of animals independently of the rest of the genome. This would be facilitated by the existence of genomic libraries. The screening of a genomic library might also provide the basis for a decision to thaw and transfer embryos from a specific population for further study as live animals.

RECOMMENDATIONS

With few exceptions, mostly in developed countries, little is known about breeds that have become extinct. Similarly, for most endangered stocks, no objective assessment has been made of their relative merits or unusual qualities. In many cases there may not be time to characterize endangered breeds before they disappear. Although loss

Scottish Highland cattle are very hardy and withstand long periods of cold, wet weather. Credit: Cotswold Farm Park.

of breeds suggests a decline in genetic diversity, the extent of actual losses is not understood. Nevertheless, the general concern is that some unique and potentially useful germplasm is vanishing—germplasm that at some point could play a valuable and necessary role in increasing the efficiency of livestock production (Alderson, 1990a; Wiener, 1990). The committee considered assuring the availability of these resources to meet unseen needs as a strong argument in support of conservation efforts, especially in view of the contributions of landraces in improving agricultural crops in both developed and developing countries (National Research Council, in press).

Sampling strategies for populations must be structured to prevent inadvertent loss of genes through the use of improper methods or an inadequate number of samples.

The development of collection strategies must include an assessment of the size of a population, its genetic structure, and its geographic distributions. This information must be available to scientists and policymakers who are developing collection strategies. Given that sampling 25 males and 50 females will reduce loss in genetic variation at any one locus to less than 1 percent, the committee concluded that the following standards for population sampling should be adopted: for semen, 25 sires, each sufficient for 200 to 500 inseminations; for embryos, 25 sires and 50 dams, each with 10 to 25 embryos; for live animals, 25 breeding sires and 50 breeding dams.

Private efforts to preserve animals in breeding herds should be encouraged.

Private individuals and interest groups, typically in developed countries, maintain a wide array of livestock for reasons not necessarily linked to genetic conservation. These efforts, however, can provide an important supplement to public efforts to manage herds and flocks. The RBST and the AMBC are examples of successful private efforts. The presence and activities of such groups enable a broader diversity of populations to be preserved than would otherwise be possible with limited resources. The degree of government involvement in private conservation efforts may vary from exchange of information to direct subsidy of private conservation efforts.

The preservation and management of endangered and unique populations as breeding herds or flocks should be supplemented by cryopreservation of their germplasm.

Cryogenic technologies are an important supplement to maintaining live herds. However, many region-specific populations in the most immediate danger of extinction may not be from widely studied breeds (Wiener, 1990). It may be difficult to save some of these populations using current cryogenic technology. Research and development are also needed to refine field techniques and to evaluate within- and between-breed responses to collection and the viability of semen and embryos after cryogenic storage.

3

Measurement and Use of Genetic Variation

A large part of observed variation in livestock species is genetic—that is, it arises from differences in the genes carried by various individuals. Genetic variation can be quantified at several levels: among species, among major types within a species, among breeds within a major type, between breeders' lines within a breed, and among individuals. Thus, for example, differences between sheep and goats (species), between hair sheep and wool sheep (types), between Dorset and Merino sheep (breeds), between commercial stocks or strains of Merino sheep (breeders' lines), and between individuals of the Merino breed can all be attributed to variation at the level of the gene. The term *genetic diversity* is used to express the degree of this variation.

In animal agriculture, manipulation of genetic variation by controlling reproduction (through selection and crossbreeding) has been the foundation for improving livestock populations. The responsiveness of a population to artificial selection and to changes in environment or selection goals ultimately depends on a reservoir of genetic variation.

THE INFLUENCE OF HUMAN SOCIETY

Domestic animal species in almost all cases do not exist as single, uniform breeds. Instead, their association with humans has encouraged the development of a highly complex overall population structure and the existence of many different crossbreeds. In many cases crossbreeds have developed fortuitously as a by-product of the migration of human populations; in others, they have developed as a result of human preferences for certain types of animals.

The migration of human populations across the globe resulted in a parallel migration of their domestic animals, and it encouraged the introduction of domestic species into regions they would not likely have colonized successfully without human assistance. Once a species was introduced, the process of selective adaptation to the new environment began, and in many cases it was furthered by isolation of the newly established population from its parent population. Over time, the joint effects of selection and isolation led to populations that gradually became genetically distinct from other populations maintained in different environments. For each domestic species the geographic centers of initial domestication can be identified, and the distribution of breeding populations about those centers, with attendant adaptation to different environments, can be traced.

Human needs and preferences also encouraged more varied populations of domestic species. The development of wool sheep, of dairy cattle and goats, and of chickens and ducks with enhanced egg-laying capacity provides examples of domestic populations bred to meet specific needs. When such developments are coupled with preferences for different color patterns or morphological types, the result is another rich source of differentiation among livestock populations.

Subdivision into Breeds

The subdivision of domestic species in response to different human preferences reached its zenith in the 1800s with the formation of breed societies in western Europe and in the regions colonized by western Europeans. The concept of breed identity rested on the idea that uniformity of type was an important goal in livestock production and could only be achieved by controlled matings involving animals of known parentage. Thus, initial entry into breed society records was restricted on the basis of color or morphological type. Continued recognition of breed membership required documentation of ancestry. The power of breed isolation can best be seen in the tremendous genetic diversity in horses and dogs; breed isolation also brought forth the rich array of livestock in western Europe.

Breed designation is primarily a western European and North American concept, but it is often applied to any reasonably identifiable livestock population. In reality, breed designations cannot be accurately used in much of the developing world, where pedigreed adherence to breed identity is not maintained, and controlled mating is not usu-

A chicken flock of mixed origin are fed outside a henhouse in Burkina Faso. Credit: Food and Agriculture Organization of the United Nations.

ally followed. In these countries, distinct, identifiable populations are likely the result of geographic isolation or phenotypic selection.

Even in developed countries the designation of breed identity does not carry with it the unambiguous distinction of genetic uniqueness that is often assumed. Breeds have historically been combined with, or absorbed by, other breeds to obtain rearrangements of the gene pool. Specific genes are not lost in this process; they are just transferred into new populations. In the U.S. beef cattle herd, for example, many animals express apparent phenotypic breed identity (usually as a result of color), but only about 5 percent of the total are registered and possess pedigree-documented breed membership (Hawkins, 1988). Thus, the genes and gene combinations carried by most breeds almost certainly also exist in the overall cattle population. An advantage of breed identity is the concentration of specific genes at high frequency and the ability to locate them reliably within those populations. In contrast, selection and screening would be required to extract the same genes from a highly heterogeneous cattle population.

The crossing of breeds to obtain more heterogeneous populations does not, in itself, lead to reduced genetic diversity. The genes from the combined populations should still exist within the overall population, although they could subsequently be lost through selection, continued crossing back to one of the parent breeds, or random chance if resulting populations are small. The introduction of imported stocks into a nation is often viewed as a source of genetic diversity that can subsequently be used in population improvement.

WHAT IS A LIVESTOCK BREED?

The term *breed* as a formal designation often has little meaning outside areas of Western influence, where pedigree recording is often nonexistent. For the purposes of this report, the term *breed* means any recognizable interbreeding population within a livestock species. A degree of mixing may occur among different breeds where mating is not strictly controlled. However, recognizable, regional stocks or populations exist worldwide that are analogous to Western breeds.

The concept of a livestock breed, in which all members of the population have a pedigree tracing their ancestry to animals of the same breed, was developed primarily in Western Europe during the eighteenth century. In its strictest sense, a breed designates a closed population—mating pairs are drawn only from within the population and relationships among individuals are documented by recorded pedigrees for all animals. Its members share certain recognizable phenotypic characteristics, such as color, horn shape, and body type, that are a feature of their breed.

Members of a breed share a common ancestry and selection history. For example, the Holstein breed of dairy cattle, developed from the Dutch Black Pied breed, have been selected to produce large quantities of relatively low-fat milk, while the Jersey breed of dairy cattle, from the Channel Islands, has been selected to produce smaller quantities of richer, high-fat milk.

In general, the term *breeding population* refers to an interbreeding group with some identifiable common appearance, performance, ancestry, selection history, or other feature. Without good measures of genetic distance between groups, ad hoc judgments must be made about which individuals constitute a breeding population. With a good sampling procedure and by keeping individual samples separate, however, reclassification can be done retrospectively if better information accrues.

Most livestock populations are outbreeding, so individuals chosen at random should be representative of the population and of its genetic variation. What may be uncertain is the extent of gene flow between different subpopulations, such as geographic groups or breed subtypes.

With breed preservation, however, the packaging as well as the overall maintenance of genetic diversity are important. Genes and gene combinations are usually more readily accessible in relatively fixed populations than in large, heterogeneous populations. In milk production in cattle, for example, the Holstein breed is widely recognized as superior in terms of total volume of milk produced per lactation. Although the genes controlling lactation clearly exist in other breeds and, over time, could be identified and their frequency increased to yield a genetic capacity for high milk production, a similar result could be obtained much more rapidly by crossing other breeds with the Holstein to achieve a rapid adjustment in the frequency of alleles conducive to high levels of milk production. A strong argument for breed preservation, then, is the need not just to maintain genetic diversity but also to maintain its accessibility in predictable source populations.

FACTORS AFFECTING GENETIC VARIATION

Allelic frequencies are altered by four forces: selection, migration or gene flow, mutation, and random genetic drift. Of the four, mutation is the only force that produces new genetic variation (in the form of new alleles). The other forces may alter the genetic composition or structure of a population.

Selection

Selection can be either natural or artificial (imposed by humans); both are strong forces that alter the genetic composition of many livestock populations. Natural selection is a continuous process reflecting environmental pressures, and occurs when a certain genotype confers a reproductive advantage on an individual relative to other genotypes in that population. Artificial selection is superimposed on natural selection by animal breeders in an effort to change various traits in preferred stocks. The rate of response to selection of either kind depends on the variation in the population. More variable populations are expected to produce more extreme types that can be selected or identified and propagated to produce more rapid genetic changes.

In livestock improvement, selection has operated historically at the level of the phenotype. In recent decades, however, methods such as progeny testing have been developed to estimate the genetic merit of an individual based on the performance of its offspring. Now, molecular methods for discerning the genotype are opening up new possibilities for artificial selection based on the actual genes carried by an individual.

Natural selection applies selection pressure on a population for certain characteristics, and it may or may not be antagonistic to artificial selection. Likewise, environments vary widely, and a genotype that is at a selective advantage in one environment may lose that advantage or actually be at a disadvantage under different conditions. For example, the small goats living in the tsetse fly zones of East and West Africa are much more resistant to trypanosomiasis than imported European breeds. This phenomenon eventually leads to the development of genetically distinct populations adapted to specific environments. Genotype-by-environment interactions maintain allelic diversity if subpopulations of a species are raised under a wide range of production systems.

Breeding goals have changed over the years, and artificial selection practiced by animal breeders has varied accordingly. Currently, a range of selection goals can be identified for almost all livestock species; the most pronounced differences are found for ruminants raised for their meat. In chickens separate breeds and crossbreeds have been selected for eggs (layers) and meat (broilers). Even within the highly selected lines derived from these commercial breeds, goals have not remained constant over time. In broilers, for example, the emphasis has moved away from birds with faster growth rates to those with more desirable body composition or those that use feed more efficiently.

Migration

Migration or gene flow occurs as a result of the movement of individuals between populations. In livestock production, crossbreeding (mating among two or more different breeds) and backcrossing (mating the crossbred progeny back to one of the parental breeds) are used to move favorable alleles from one population into a second population and can result in extremely rapid changes in gene frequencies. Crossbreeding, however, can break up favorable combinations of alleles that can only be reconstituted through further breeding and selection.

Mutation

Mutation refers to a structural change in a gene, and it is thought to be a random occurrence. The probability of improving a gene function as the result of a mutation is extremely small. The genetic variation that has accumulated over many thousands of generations, however, reflects not only the environmental conditions to which that population was exposed but also the mutations that have been retained in the gene pool.

An important consideration in conserving genetic diversity relates to the extent to which mutation is a factor in maintaining and creating genetic variation over time. Selection responses in livestock populations may continue in part because of genetic variation contributed by mutation (Enfield, 1988; Hill, 1982; Hill and Keightley, 1988). Testing the value of new or newly recognized mutations, however, can be time consuming. In contrast, previously existing mutations in the population presumably have already been studied and may be of greater value as a genetic resource.

Genetic Drift

Genetic drift refers to the random changes in allele frequency. When population sizes are small, random genetic drift can lead to loss of alleles owing simply to chance, particularly if they are present at low frequencies. A practical consequence is that the number of breeding individuals maintained in a preserved population must be sufficiently large to avoid the potential for decreased genetic variation resulting from genetic drift. Guidelines for the number of animals required in breeding populations to minimize this risk have been discussed (Smith, 1984b).

GERMPLASM USE

Management programs for genetic resources are influenced by the ways in which genetic diversity is used. In general, germplasm use follows one of two patterns. The first pattern pertains to programs for industrialized stocks of poultry, pigs, and dairy cattle, which occur primarily in developed countries but which are also being aggressively marketed for use in developing nations. The second is more appropriate for grazing ruminants worldwide and is generally applicable to any stock maintained under extensive conditions.

Industrial Stocks

The development of industrialized stocks of poultry, pigs, and dairy cattle has been predicated on the ability to achieve uniform, high-quality conditions in a production environment. When achieved, this environmental stability means that minimal selection effort is

Genetically improved *Bos taurus* cattle, such as these Hereford in Kenya, are particularly desired by farmers because they are more productive than indigenous breeds. However, highly productive grade cattle are also susceptible to tick-borne diseases, such as theileriosis, a debilitating and often fatal disease caused by a protozoan parasite. Sprays and dips to keep the animals free of ticks and drugs to control the parasite in infected animals are costly and difficult to administer. Credit: International Laboratory for Research on Animal Diseases.

necessary for environmental adaptation. It also facilitates the development of specialized stocks of very high genetic merit for specific production traits, such as egg production in layer strains of poultry or milk production in Holstein cattle.

Breeders of the industrialized livestock populations have tended to move toward development of one (in the case of Holstein dairy cattle) or a few (in the case of poultry) highly selected strains that have come to dominate production. Depending on the fecundity of the species, purebred lines (in dairy cattle) or specific crossing of specialized sire and dam lines (in poultry) may be used. A relatively wide array of germplasm may have been sampled, or at least comparatively evaluated, in the early formation of the populations. Current emphasis, however, is on selection within existing industrialized stocks; the focus is normally on a small set of traits, and the goal is to achieve maximum rates of genetic change. Selection has been effective, and the industrialized stocks are now clearly differentiated from the original stocks.

The possibility of drastic changes in production and marketing conditions, and the need to return to, or sample genes from, preindustrialized stocks, cannot be totally discounted. More pertinent, however, is the fact that the large differences in production between industrial and nonindustrial stocks will likely require that the capture and use of genes residing in nonindustrialized stocks of livestock species be achieved through marker-assisted selection and, perhaps, molecular isolation and transfer of individual genes independent of the remainder of the genome. These technologies are being developed now (see Chapter 4), and they provide strong justification for the preservation and characterization of the genetic resources of poultry, pigs, and cattle.

Nonindustrial Populations

For extensively managed ruminant species raised for meat and fiber, and for nonindustrialized populations in general, germplasm

Introduced animals with superior production characteristics are often crossbred with indigenous breeds to confer disease resistance and adaptation to environmental extremes. Pictured is a cross between the Landim landrace of Mozambique and a breed of the Alpine group of dairy goats. Credit: International Livestock Center for Africa.

use differs markedly from that for industrialized stocks. In extensively managed populations the ability of individual animals to interact effectively with their environment, to forage for feed, and to deal with climatic stress is critical. Generally, the efficiency of grazing ruminants can be increased by improving their ability to adapt to their environment.

The existence of many distinct populations, each adapted to a particular environmental niche, can best ensure the capability to improve meat and fiber ruminant species. Because many of these niches occur repeatedly around the globe, several regional populations adapted to similar niches may exist at any given time. Environmental stabil-

THE PACKAGING OF GENETIC INFORMATION

A gene, in the classical sense, is the basic unit of heredity, and it has one or more specific effects on the organism. Genes are segments of DNA (deoxyribonucleic acid) in the nucleus of the cell, linearly arranged to form threadlike structures called chromosomes. The term *locus* refers to the position of a gene on the chromosome, and it is sometimes used interchangeably with the term *gene* to refer to particular segments of DNA that influence a trait.

Alternate forms of a gene found at the same locus are called alleles. Some genes have many alleles, while for others only a single allele is known. Multiple alleles result in multiple phenotypes for a particular trait; that is, the trait will vary depending on which allele is present at the locus. Multiple alleles, for example, have been identified for many of the genes that code for blood proteins. The allelic frequency refers to the proportion of loci in the population as a whole occupied by each allele. A gene for which there is more than one allele is called polymorphic.

Livestock animals are diploid—that is, they carry two copies of each gene. If both copies are the same allele, the individual is said to be homozygous; if the two copies are different alleles of the gene, the individual is heterozygous for that locus. An individual's genetic composition or genotype, in conjunction with the environment in which that individual is found, determines the phenotype or observable characteristics.

DNA is the chemical that carries genetic information in almost all living organisms. The DNA molecule itself is an elongated double helix, often compared to a long, twisted ladder. Corresponding to the rungs of the ladder are two bases, a base pair. The sequence of base pairs at a given locus confers the specificity required for transmitting information by means of the genes. Owing to variability of the base ity, however, is rapidly falling victim to human technology. Environmental niches can change rapidly as conditions are improved or degraded. The continued adaptation of grazing ruminants to these changing environments can be achieved by selection, but it is often accomplished more efficiently by recombining several existing populations to form a new germplasm base consistent with the new production conditions. Thus, Sahiwal cattle from the Indian subcontinent have contributed to milk-producing stocks in Africa and Australasia; prolific Finnish Landrace sheep have contributed to the development of several new breeds worldwide—for example, the Polypay sheep for use in improved grazing areas of North America; and Africander

---

pairs, the number of different alleles that could theoretically be formed from even a short piece of DNA is extremely large. For instance, a segment of DNA with only 10 base pairs could have over a million different codes. Genes average about 1,000 base pairs, but some are in excess of 100,000 base pairs; thus, the DNA structure provides for an amazing amount of variation. The difference between two alleles is often as simple as a substitution of a single base pair, but that may correspond to a significant difference in the phenotype resulting from those alleles.

Certain traits are controlled by a single gene; these are referred to as qualitative or Mendelian traits. Red versus black coat color in cattle, for example, is controlled by a single locus. Many characteristics, however, are influenced by a larger number of genes; these are called quantitative or polygenic traits. The cumulative action of these genes influences the expression of the trait, but the effect of any single gene is small and cannot generally be isolated in the phenotype. Important production traits, such as rate of weight gain, milk yield, and litter size, are examples of polygenic traits. Occasionally, a major gene, one having a stronger influence on the trait, can be detected, but there is nevertheless modification of the trait by other loci that have smaller effects.

An understanding of the extent of allelic or genetic variation is a key factor in conservation genetics. For some loci it is possible to estimate allelic frequencies for a group of animals, such as a breed. The number of loci for which allelic frequencies have been quantified, however, is but a handful of the estimated 50,000 to 100,000 genes operative in most mammalian genomes. Alleles contributing to polygenic traits cannot generally be identified, and measurements of their frequencies have not been feasible. Allelic frequencies undoubtedly vary among subpopulations within a species, but the extent and significance of that variation are not usually known.

cattle from southern Africa have become a part of the Barzona breed for use in the arid regions of the United States.

RECOMMENDATIONS

Globally, a vast amount of diversity exists: among habitats or ecosystems, among the species found within an ecosystem, and among the individuals of a given species (Office of Technology Assessment, 1987; Wilson, 1988). Variation among individuals of a species leads to differences in their ability to respond to the environment, to differential reproductive success, and, ultimately, to continued evolution and adaptation of the species.

The status of unique populations should be monitored carefully, and when necessary, appropriate action should be taken to prevent loss.

Rich diversity exists within domestic species and reflects genetic differences among the populations. Breed preservation protects genetic diversity and enables more rapid use of conserved genes. Loss of unique breeds or populations implies loss of unique alleles or allelic combinations.

Existing breeds and populations should be evaluated to the extent possible to assess their potential to contribute to livestock improvement.

The genetic characterization of large numbers of indigenous breeds is costly and time consuming, but the evaluation of promising breeds is a critical part of livestock improvement efforts. The general use of unimproved or undeveloped germplasm as a source of genes for improvement is less likely for modern industrialized and intensively managed livestock than it is for improvement of nonindustrialized, extensively managed livestock, such as the grazing ruminants. However, the possible value of specific genes from unimproved breeds that are adapted to certain environments is a strong future possibility. For example, genes for disease resistance or ability to survive with less water may be incorporated into intensively managed breeds to adapt them to harsher environments.

The degree of differentiation between the industrialized stocks and other populations of the same species has now progressed to the point that use of genetic material from nonindustrialized or preserved stock is no longer considered attractive in conventional breeding methods, barring major changes in the production system. However, useful individual genes or gene complexes, such as those that confer disease resistance, exist in nonindustrialized stocks. The cost of inserting those genes by traditional crossing and backcrossing techniques is considered prohibitive. Advances in molecular biology may foster, through gene transfer, the use of unique genes from unimproved stocks.

In grazing ruminants, it is possible that a number of existing breeds may have much to contribute to current commercial stocks, either in crosses or as purebred lines. The evaluation of these breeds, then, becomes a priority. The potential of preserved breeds to contribute to future production systems is also greater for the grazing ruminants because the global range in production environments encompasses such a wide range of conditions and because highly specialized, industrialized stocks do not yet exist for these species.

4

New Technology and Its Impact on Conservation

New technologies are creating greater possibilities for preserving and using animal genetic resources. New molecular genetic technologies can provide valuable tools for characterizing and using genetic stocks. Of particular importance will be improved capacities to search preserved germplasm for useful genes that can be incorporated into existing breeding stocks. Improved methods for diagnosing and controlling diseases will significantly reduce barriers to exchanging germplasm across borders and thereby enhance the value of preserved stocks for future use.

This chapter describes new and emerging technologies to quantify genetic variation, to transfer genes from one animal to another, and to optimize reproduction. With further development, many of these technologies could provide powerful tools for preserving and using livestock genetic resources. The chapter ends with a brief discussion of the impact of health status on the movement and use of genetic resources that is followed by a set of recommendations.

METHODS TO QUANTIFY GENETIC VARIATION

Biotechnological methods that enable scientists to characterize an animal's genes and the gene pools of breeding populations are becoming a reality. These methods will aid in making decisions about the uniqueness of germplasm for preservation. The study of the structure and function of genes at the molecular level (molecular characterization) in a breeding population can help determine the similarity of the genetic material carried by two or more populations and the genetic variation within a population. The identification of individual genes, on the other hand, will allow the genetic makeup of an animal (genotype), as distinguished from its appearance (phenotype), to govern the process of choosing the parents of the next generation (artificial selection). If genes for production traits can be identified, for example, they can be selected for even if they are not expressed in an individual animal. Alternatively, if they can be linked to genes with known locations on chromosomes and clear-cut phenotypes (marker loci), selection can be done using these points of reference. This allows selection for a desired character even though the genetic basis for the desired phenotypes may not be known.

Molecular Quantification of Genetic Diversity and Distances

Several techniques are used to estimate genetic differences between populations. Each has its own set of assumptions and applicability to a given problem. All attempt to quantify genetic distance between populations, which is a measure of the number of genetic differences between two populations or species. Molecular techniques for estimating differences have been reviewed by Chambers and Bayless (1983), Nei (1987a), and Sharp (1987). These methods include DNA (deoxyribonucleic acid) sequencing, DNA-DNA hybridization, protein electrophoresis, immunologic methods, and restriction fragment length polymorphisms (RFLPs).

DNA sequencing involves determining the sequence of base pairs in sections of the DNA. Although techniques for automating DNA sequencing are being developed and improved, the process is still too expensive and time consuming to be routinely applied. A reliable estimate of sequence divergence between two populations would require sequencing of many genes for many animals within each population. This is not currently feasible in animals, but it could become the main method for determining genetic distance in the future.

DNA-DNA hybridization requires the isolation of DNA from two related genomes. The DNA is treated with heat, causing the DNA double-stranded helix to disassociate and become two single-strand DNA molecules. When the single-strand DNA from two different sources is combined, it will re-form into double-stranded DNA upon cooling. The stability of the re-formed double-stranded DNA depends on the exactness of the match. The extent of nucleotide differences between the two different strands can then be approximated upon reheating and measuring the temperature at which the double strand dissociates again into single strands; this provides an index of relatedness and so, conversely, of genetic distances.

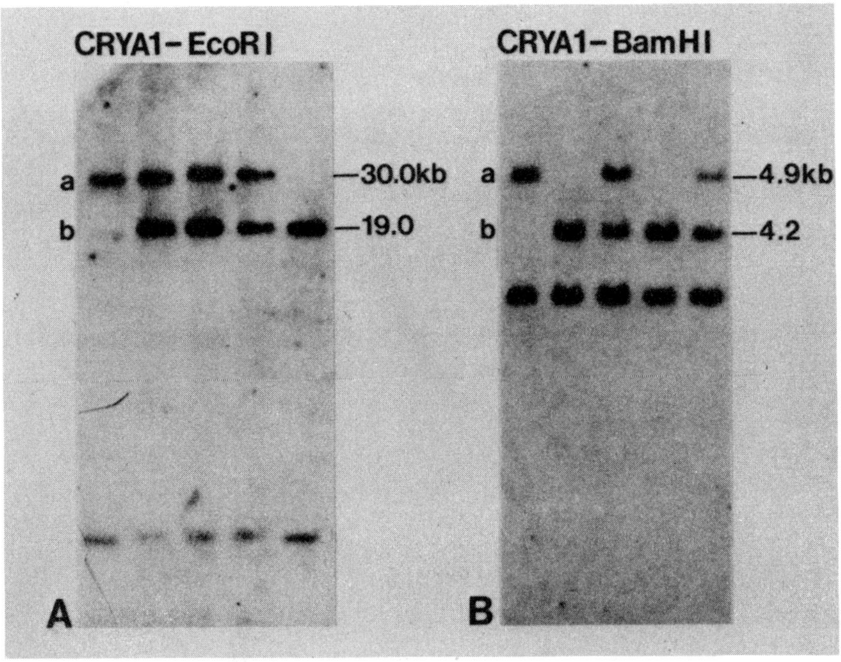

Restriction fragment length polymorphisms from five individuals in a population of crossbred cattle are revealed using two different restriction enzymes (EcoRI and BamHI) and a probe for the eye lens protein, α crystallin (CRYA1). Alleles that are polymorphic are labeled a or b with each fragment size expressed in kilobases (kb). In this example, a single probe and two restriction enzymes help distinguish four sites of genetic difference among the five individuals. Source: Thielman, J. L., L. S. Skow, J. F. Baker, and J. E. Womack. 1989. Restriction fragment length polymorphisms for growth hormone, prolactin, osteonectin, α crystallin, γ crystallin, fibronectin and 21-steroid hydroxylase in cattle. Anim. Genet. 20:257–266. Reprinted with permission, ©1989 by Blackwell Scientific.

O'Brien et al. (1985) applied this technique to estimate the genetic distance between species of higher primates and carnivores and compared it with more classical estimates derived from protein variation (isozyme) analysis. Among the methods for estimating molecular distance, DNA-DNA hybridization produces data that more closely agree with those derived from evolutionary studies of primates. Genetic distances between these primates as measured by isozyme analysis are similar to those observed between cattle breeds (Sharp, 1987). Thus, DNA-DNA hybridization might be useful for quantifying relationships among breeds of livestock.

Electrophoretic analysis of protein variation has been the most popular technique for estimating genetic distances. It has been widely used for many years to detect genetic variation in populations at the molecular level (Harris, 1966; Lewontin and Hubby, 1966). The method is based on the principle that proteins of differing sizes and structures will have differing net charges, and thus differing mobility, when placed on a gel and allowed to move in response to an electric field. The proteins detected by electrophoresis are various enzymes, and a small amount of blood or tissue is a sufficient sample for measurement.

In typical outbreeding mammalian populations, two or more alleles were found at 15 to 25 percent of the gene loci examined (that is, they were polymorphic). The average heterozygosity in individuals over all loci is between 3 and 10 percent (Baccus et al., 1983; Nevo, 1978; Selander and Kaufman, 1973). To date, about 30 polymorphic loci have been identified that can be used to analyze the genetic structure of livestock species.

The identification of genetic variation by electrophoretic analysis underestimates the genetic variation at the DNA level. Only a small fraction of the genome codes for proteins, and thus, a random sample of the DNA is not examined. Even in the coding regions, sequence changes in DNA do not always lead to amino acid changes; thus, they will not be reflected as electrophoretic changes. Even if changes do occur, only those changes that affect the net charge of the protein will be detected. Nevertheless, sufficient protein variation can be found to quantify genetic distances (Nei, 1987a), and electrophoresis has been successfully applied to the study of genetic distances between breeds (Glodek, 1982; Gonzalez et al., 1987; Kidd et al., 1980).

The detection of allelic variation in blood group proteins by immunological techniques (Rendel, 1958; Stormont, 1959) is comparable to protein electrophoresis in application and limitations. The number of identified variant proteins, however, is less. For example, only about 12 blood proteins have been identified in cattle (Standing Committee on Cattle Blood Groups, 1985). Detection of variation by other methods, such as the use of molecular markers, appears to have greater potential (Baker and Manwell, 1991).

A relatively new method, RFLP analysis, can be used to characterize populations. Restriction endonucleases are enzymes that cut DNA at specific sequences, called restriction sites (Zabeau and Roberts, 1979). When genomic DNA is cut with one of these enzymes (more than 100 are commercially available), a mixture of DNA fragments of varying sizes results; these are called restriction fragments. Electrophoresis is used to separate the restriction fragments according to size. The fragments are then transferred to a filter by blotting (Southern, 1975), which produces a replica of the gel on which DNA-

DNA hybridization can occur. Specific DNA sequences, known as DNA probes that are labeled with a radioactive isotope of phosphorus (^{32}P), bind to complementary DNA fragments on the filter. Positions of the probes, which correspond to specific restriction fragments, can be determined using autoradiography.

If, in two DNA samples, fragments of differing size result from the use of the same restriction enzyme, they reflect a polymorphism or difference in the base-pair sequence in the restriction site (that is, the site at which the enzyme cut the DNA). Such a polymorphism in a restriction site is evidenced by bands at different positions on the filter, because electrophoretic mobility is a function of fragment size. Assuming that the DNA of domestic animals proves to be as polymorphic as that of humans, a polymorphic site should occur in one of every 100 base pairs (Jeffreys, 1979).

The RFLP technology and its application to animal breeding were reviewed by Soller and Beckmann (1983). The authors summarized data from humans, mice, and cattle and estimated that 5 to 10 percent of enzyme-probe combinations can be expected to reveal an RFLP. Consequently the level of genetic variation revealed by RFLP analysis far exceeds that from protein electrophoresis. Because the technique is relatively new, very few applications have yet been made to estimating genetic distance. Mathematical models have been developed, however, for estimating genetic divergence from restriction site polymorphisms (Nei, 1987b; Nei and Tajima, 1981). Wainscoat et al. (1986) have applied these models to studies of relationships among human populations.

As procedures for evaluating genetic distance between populations are refined, they will provide much-needed information on the distribution of genetic variation within domestic species (Theilmann et al., 1989). This information will greatly assist in making conservation decisions by allowing more quantitative evaluation of the probability that an endangered population contains alleles that are unique to the species. Studies to characterize the genetic structure of global livestock breeds should thus receive high priority.

Characterizing Mitochondrial DNA

In cells, DNA is also found outside the nucleus in small oblate bodies, the mitochondria. Mitochondrial DNA (mtDNA) is considerably smaller than nuclear DNA and is easily isolated. Consequently, mtDNA has been well characterized in most animals and is widely used in evolutionary studies. Mitochondrial DNA genes are transmitted through the female parent and so reveal only the maternal lineage. This limits their usefulness in fully defining breeding populations of livestock. The mitochondrial genomes of several livestock species, including cattle, have been sequenced.

Developing Genetic Maps

Gene mapping assigns a locus to a specific chromosome or determines the sequence of genes and their positions relative to one another. Knowledge of RFLPs has greatly aided gene mapping (Beckmann and Soller, 1983). A genetic map saturated with RFLP markers at frequent intervals would ensure that an association could be detected between a marker and any important gene. No agriculturally important gene would be beyond the reach of marker-assisted selection. Individuals with desirable genes could be selected on the basis of markers even in the absence of phenotypic expression of the gene or of knowledge of its molecular basis. Rams carrying the Booroola gene for prolificacy, for example, could be identified without progeny testing their daughters if the gene were mapped near a marker.

RFLPs appear to be widespread in all species, although only a small number have been defined in livestock to date (Foreman and Womack, 1989; Fries et al., 1989). Applications of RFLP-assisted mapping to human genetics are reviewed by Watkins (1988). More than 4,500 loci have been cataloged for the human genome, and the number is rapidly increasing (Human Gene Mapping 10, 1989). The power of this technique for genomic analysis of livestock has only recently been recognized. Recent literature suggests a trend to exploit these markers, at least in cattle, for which more than 30 have been described to date (Beckmann et al., 1986; Fries et al., 1989; Hallerman et al., 1988; LeQuarré et al., 1987; Theilmann et al., 1989).

An even more powerful class of RFLPs, called variable number of tandem repeats (VNTRs), has contributed much to expanding the human gene map (Nakamura et al., 1987). Searches for VNTRs in livestock are only beginning. The advantage of these markers is in their extreme variability; they exist as multiple alleles with a high probability of being heterozygous in any given animal. Single-locus VNTR probes have not been described for livestock, but the extensive restriction fragment variation associated with these probes has been described (Vassart et al., 1987). The VNTR probes may be particularly useful in establishing parentage because they will be more similar in closely related animals.

Linkage groups refer to loci (genes or markers) that are in close proximity on the same chromosome and, therefore, are generally inherited together. Genetic mapping of a linkage group (linkage mapping) requires studying inheritance of the markers or phenotypes of interest over several generations. Genetic mapping is easier when multigenerational pedigrees of large families are available. Several linkage studies have been performed in livestock based on the small battery of markers provided by enzyme and blood group research (Andersson, 1986; Hines et al., 1969; Larsen, 1977). These studies can now be replicated using the much larger number of gene markers provided by RFLP analyses to detect many more linkage groups.

Physical mapping, which determines the precise location of a gene on a particular chromosome, can be accomplished in livestock species with the same technologies used for humans (Fries et al., 1988). Mammalian chromosomes are sufficiently preserved across species to allow the identification of homologous (similar) segments between such diverse species as cattle and humans (Womack, 1987; Womack and Moll, 1986). Their identification permits extrapolation of mapping data from one species to another. This capability may be particularly important to animal geneticists in light of the wealth of data being obtained from human genome mapping.

The current activity in animal gene mapping suggests that useful maps for most of the livestock species will be developed by the mid-1990s. In 1987, 50 genes had been mapped in cattle, 31 in sheep, 38 in pigs, and 21 in horses (Lalley et al., 1987). By 1989, the number of genes mapped in cattle had already doubled. An extensive linkage map of the domestic chicken was published in 1987 (Somes, 1987).

DNA LIBRARIES AND GENE TRANSFER METHODOLOGIES

Gene transfer involves identifying and isolating a gene from one genome and transferring it into the genome of another individual. Transfers may be within the same species or from one species to another. To date, most work on mammals has been with mice; however, application of gene transfer technologies in livestock species, with the objective of improving production traits, is receiving greater attention (First, 1991; Gibson and Smith, 1989; Pursel et al., 1989; Smith et al., 1987; Wagner, 1986). A closely related technology is the formation of DNA libraries, which has important potential for preserving germplasm. Genomic DNA and DNA libraries from different genetic sources also represent valuable research resources.

Establishing an Animal Genomic Library

An animal genomic library is a collection of cloned fragments representative of individual genomes. In a genomic library, genes from an animal, in the form of segments of DNA, are incorporated into the genome of a simple organism, such as a bacteriophage. As the "carrier" organism multiplies, it duplicates its genome as well as the added segments. The collection of cloned DNA segments that comprise the genomic library, however, has no catalog; additional screening is needed to determine which genes are present and where they are located within the library.

Cloning Genes in a Library

A genomic library can be established by isolating DNA from virtually any tissue (for example, organs, blood, semen). The DNA is cut into small segments with specific DNA-cleaving enzymes (restriction endonucleases). Random DNA segments are then inserted into the intact genome of a virus (such as bacteriophage lambda), which can be rapidly reproduced in its bacterial host, *Escherichia coli*. Many copies of the bacteriophage lambda are made within each bacterium to yield multiple cloned copies of the inserted DNA segment. DNA segments, containing one or more genes, are inserted into separate lambda genomes. These bacteriophages can then be propagated separately as clones derived from the single "transgenic" (because it now contains genetic material transferred from the source animal tissue) lambda phage.

The complete genomic library consists of a suspension of millions of transgenic lambda phage organisms, each carrying a gene or genes from the source animal. This suspension may be cryopreserved indefinitely. It may also be multiplied to produce additional copies of the library. When an appropriate volume of the transgenic bacteriophage suspension is placed on a plate containing a layer of *E. coli*, each individual transgenic organism will form a separate colony (plaque). In theory each colony contains a piece of the original animal DNA and can be isolated for further growth and testing. These colonies represent individual "cloned" segments of DNA.

Screening an Animal Genomic Library

Screening a genomic library is the process of discovering which colony contains a particular DNA segment. All phages grown from that colony will contain the gene of interest.

The screening of a genomic library requires a means of identifying the appropriate colony. This is accomplished by using a radioactively labeled DNA probe. A messenger RNA or segment of a known gene can be used to construct a DNA probe that has a nucleotide base sequence complementary to all or part of the DNA sequence of interest. When placed on the growth plate containing the entire library, the probe will bind only to colonies containing the complementary DNA segment. If the plate containing the library is then placed over X-ray film, a dark spot will develop under that colony, thereby identifying it.

Once a colony has been identified and isolated, large quantities of the organisms can be grown to increase the number of copies of its particular DNA segment. The DNA sequences can then be separated from the virus DNA using the same endonuclease originally employed in constructing the library. To isolate a single gene it may be necessary to obtain DNA segments from more than one colony because intact genes are rarely found in a single clone. Using this technique, individual genes can be located, sequenced, studied, and used for gene transfer.

Transfer of a Cloned Gene into Another Animal Line

A cloned gene from a genomic library may be incorporated into the genome of another animal. In mammals the cloned DNA segment is microinjected into the pronucleus of the fertilized egg. Enzymes for maintenance and repair of the embryonic DNA are believed to splice the added gene into a chromosome. It is thereby permanently incorporated into the genetic composition of what is now termed a *transgenic animal*. Procedures for introducing cloned genes into avian species have also been developed during the past several years (Brumbaugh, 1989; Freeman and Bumstead, 1987; Gannon et al., 1990).

The percentage of embryos that are transgenic (containing the extraneous DNA segment) following microinjection of the eggs is low, generally less than 2 percent in livestock species. Because the site of the insertion cannot be controlled, the DNA segment may be incorporated anywhere in the genome. This may impair the function of another gene within the genome. Control of the insertion site has been achieved for some cloned genes in the mouse (Capecchi, 1989).

Although successfully incorporated, some transgenic genes may not be expressed. The value of those that are expressed will depend on the site (or tissue), time, and intensity of expression. So far, no transgenic livestock have proved useful commercially, and many show serious deleterious effects of the transfer (Pursel et al., 1989). Thus, although the transfer of genes as DNA from genomic libraries has been shown to be possible across species, much research and development work is needed to make it an effective tool for using animal genes.

Animal genomic libraries may also provide a means of evaluating preserved animal germplasm. If the DNA coding for specific gene products can be identified, the DNA of live or cryopreserved stored germplasm could then be screened for potentially useful variants of the gene. Specific alleles, identified by their DNA sequence, could then be bred into current stocks for evaluation.

DNA Libraries as Supplements to Conservation Programs

DNA should be considered only as an adjunct, rather than primary, means of germplasm conservation. Regeneration of a living organism is not yet possible from DNA stores. In certain situations, however, preservation of DNA stores may be a reasonable and useful supplement to other efforts. For example, a sufficient DNA sample can be obtained from blood or sperm cells collected when taking semen samples for cryopreservation. The tissue sample could be frozen without further processing or processed to yield purified DNA for storage.

Although a DNA sample can be obtained from several billion sperm cells or from 50 to 100 milliliters of blood, continuing studies of any population may require considerably greater quantities of sample DNA. This additional DNA could be derived from genomic libraries and propagated. If stores for a preserved population, such as those described above for sperm, decline, a genomic library should be established to preclude loss of potentially useful DNA.

Identifying Quantitative Traits

Techniques such as RFLP and hybridization analysis may be used to characterize the genomes of individuals in a specific population and to compare them with reference genomes from other populations. These molecular descriptors may be of importance in selecting specific samples from frozen stores. However, these techniques focus on individual genes while most production traits are affected by many genes (quantitative traits), each with individual effects that are not easily detected. Methods are being developed in animals for identifying and tagging the different segments of the genome containing genes of moderate or large effects on a quantitative trait (Soller and Beckmann, 1983). This approach has proved successful in plants (Tanksley et al., 1989).

REPRODUCTIVE TECHNOLOGIES

Until recently, genetic change within populations was, in general, relatively slow. The application of new reproductive technologies and the increased international exchange of germplasm can now enable rapid and dramatic changes. For example, in many countries, national chicken populations have changed practically overnight from genetically heterogeneous backyard fowl to selected homogeneous stocks raised under intensive conditions.

Cryopreservation of Semen and Embryos

The discovery in 1949 that semen could be protected from the harmful effects of freezing and thawing was an important development in the area of animal breeding. Considerable progress has been made since then in establishing procedures for freezing and storing the semen and embryos of a wide variety of mammalian species, usually in liquid nitrogen at −196°C. Critical elements of cryopreservation are the development of suitable media to protect cells when frozen and stored, and procedures to ensure that the rates of cooling and warming prevent the detrimental effects of intracellular freezing. A good understanding of the processes responsible for freezing injury has emerged. Cryopreservation now offers a cost-effective way to retain a sample of the genetic diversity of a population of animals in suspended animation for indefinite periods of time.

Current technology permits the cryopreservation of semen for

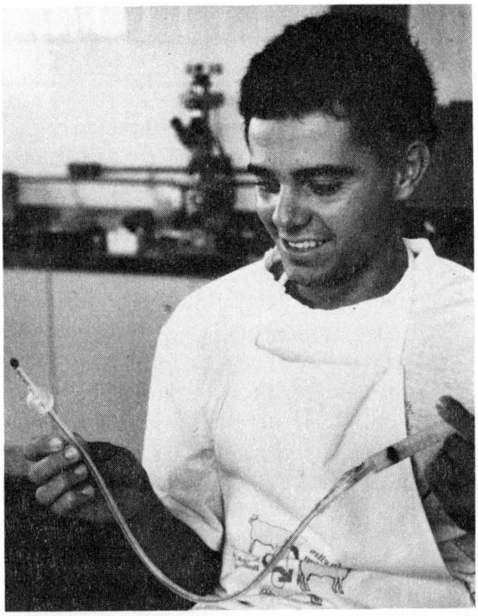

A technician in a Brazilian laboratory tests a catheter before attempting embryo collection from endangered horses in Brazil. Credit: Elizabeth Henson, Cotswold Farm Park.

livestock species. In cattle, methods for semen collection and freezing are well established, and artificial insemination is widely used, especially for dairy cattle. Cryopreserved buffalo semen is commercially available. For sheep, pigs, goats, horses, and poultry, cryopreservation methods are not yet fully satisfactory, and frozen semen is not widely used commercially. For special applications with these animals, however, cryopreservation of sperm is feasible. For

REPRODUCTIVE AND MOLECULAR TECHNOLOGIES PRESENT CONCERNS AND OPPORTUNITIES FOR CONSERVATION

Concern about conserving genetic diversity in livestock species is largely a response to technological innovations that have allowed widespread dissemination of genetic material from a limited number of highly selected breeds and individuals. Prior to the last half of the twentieth century, the breeding structure of most livestock species was nearly ideal for maintaining genetic diversity. Livestock populations were widely dispersed geographically and strongly subdivided along regional and national lines. Movement of individuals among neighboring herds was common, but the international and transcontinental movement of livestock was relatively rare, in part because of difficulties in moving large numbers of breeding animals over long distances. Fear of introducing exotic diseases also limited the movement of germplasm, as did the frequent inability of imported animals to survive and reproduce in foreign environments.

More recently, however, many of the problems of germplasm transfer have been solved, and the global movement of germplasm has greatly accelerated. The greatest impact has come from the use of artificial insemination and embryo transfer using frozen semen and embryos. The genetic makeup of national livestock populations can be encompassed by a few highly selected animals. Internationally, these technologies, coupled with improved diagnostic tests and embryo washing procedures to reduce risks of disease transmission, will soon allow individual animals to have a far-reaching global genetic influence. Techniques for in vitro oocyte maturation and fertilization and for embryo cloning promise to expand the potential impact of individual parents even further.

Although the application of these advanced reproductive technologies can be viewed as a threat to genetic diversity, they also provide powerful tools for its preservation. Germplasm from endangered breeds, unpopular lines, or populations that are being replaced by other stocks

(continued on page 90)

the minor livestock species, the efficiency of cryopreservation is either less effective or unknown (Graham et al., 1984). (See Appendix C.)

The frozen semen pool can be used in breeding live conserved stocks to mitigate the effects of natural selection, inbreeding, and genetic drift. A drawback of a semen store by itself is that purebred stock can be produced only after a series of backcrosses to current stocks. The availability of frozen embryos removes this constraint,

These twins were the third set of their type produced at the International Laboratory for Research on Animal Diseases using embryo transfer technology. A trypanotolerant N'Dama embryo and a trypanosensitive Boran embryo were implanted at the same time into this recipient Boran cow to produce hematopoietic chimeric twins. As they developed in the uterus, their blood was mixed through the placenta, which may result in each fetus taking on each other's blood characteristics. Scientists at the laboratory will challenge the twins with trypanosomes to determine which, if any, trypanotolerant traits have been transferred from the N'Dama to the Boran. Credit: International Laboratory for Research on Animal Diseases.

for then the purebred stock can be produced immediately and maintained subsequently by the more extensive pool of frozen semen. Thus the ideal package for cryopreservation is a small embryo store (100 to 200 embryos) combined with a semen store.

The first cryopreservation of mammalian embryos followed by live birth was reported in mice (Whittingham et al., 1972). Since then, embryos of about 20 mammalian species have been successfully

---

can be retained indefinitely through judicious sampling and cryopreservation of semen and embryos. The capacity to regenerate commercial stocks can also be preserved through the long-term storage of semen and embryos. Studies have amply demonstrated that cryopreservation is genetically sound and provides cost-effective insurance against loss of genetic diversity (Smith, 1984a,b). This strategy is most feasible in cattle and sheep, where use of frozen semen and embryos is most common, but it will be more difficult in swine and poultry, where embryos cannot be frozen and use of frozen semen is less common. Studies on the cryobiology of the latter species thus become a priority.

Similar arguments can be made for the potential impact of molecular biology, which allows identification of genetic diversity at the molecular level. Its impact on commercial livestock populations cannot yet be predicted. However, the possibility that future commercial stocks will contain many genetically altered individuals must be considered. Depending on the source of the introduced genetic material, these animals could represent an enhancement of genetic diversity through the introduction of novel genes to the species. It seems more likely, however, that losses in overall genetic diversity could occur if commercial populations are founded from a limited sample of genetically manipulated individuals.

Molecular technologies can also be used to enhance conservation efforts. Animals selected for conservation can be documented at the molecular level to possess desired genetic characteristics. DNA libraries, which possess cloned fragments of an animal's genome, can also preserve the genetic material of endangered stock for future use or study. However, the present inability to directly use DNA in these libraries for improving livestock means that they should not be seen as a substitute for proven methods of conservation.

Recognition of the need to conserve genetic diversity must become an integral part of genetic improvement programs. The livestock breeder and molecular geneticist must be free to continue to attempt to maximize rates of improvement in livestock species, secure in the knowledge that the same technologies that facilitate improvement efforts can also aid in securing the levels of genetic diversity required to meet unforeseen future needs.

cryopreserved, and the procedures are now routine in cattle, goats, sheep, rabbits, and mice (Leibo, 1986). Pig embryos have been successfully frozen at expanded blastocyst (entire zona pellucida) and early hatched stages, but pregnancy rates are very low (Hayashi et al., 1989). Three mouse embryo banks have been established for preserving defined inbred and mutant strains of laboratory mice (National Institutes of Health, Bethesda, Maryland; Jackson Laboratories, Bar Harbor, Maine; and Medical Research Council Laboratories, Carshalton, England).

Vitrification is another promising advance in embryo cryopreservation. The process involves rapid cooling to transform cellular liquids into a glass-like viscous solid without any ice or crystal formation. Success with embryos requires the use of a highly concentrated, yet effectively nontoxic solution of cryoprotectants as well as rapid cooling and warming rates (Rall and Fahy, 1985a,b). A vitrification procedure for bovine blastocysts has been developed by Van der Zwalmen et al. (1989).

Multiple Ovulation and Embryo Transfer

An important aspect of the use of embryos has been the development of hormonal treatments leading to greater numbers of embryos per donor female. The administration of a follicle-stimulating hormone will cause multiple ovulations. After estrus is induced, the donor female is inseminated. Embryos are usually obtained from large animals, such as cattle, buffalo, and horses, nonsurgically while those from other species are collected surgically. Yields are variable. In cattle, six to eight embryos are obtained during the procedure. Pregnancy rates of 60 to 70 percent with fresh embryos and of 40 to 50 percent with frozen embryos have been reported. These rates, however, have been achieved and are slowly improving after substantial research, development, and experience with a few breeds. Results for other breeds and species are variable, and further specific research will be required to improve methods (see Appendix B).

A new series of technologies, again mainly in cattle, involves collection and maturation of oocytes (unfertilized ova) and in vitro fertilization. Oocytes may be aspirated from the ovary of a live animal or excised from the ovary immediately after slaughter. The oocytes are then matured and fertilized in an artificial environment, and they produce normal embryos to be transplanted into recipients. These techniques are still in the developmental stage, but they will increase the potential genetic contribution of individual females to the population.

Embryo Splitting and Cloning

Cloning—the asexual reproduction of identical genotypes—has been achieved in cattle by microsurgically splitting embryos or by transferring individual nuclei derived from multicellular embryos into enucleated oocytes (oocytes with the nucleus removed). Experience in splitting and cloning embryos has been mainly with cattle and sheep (First, 1991). Small groups of identical individuals have been obtained.

Embryos can be mechanically divided to produce genetically identical embryos; the splitting is usually done at the 60-cell (morula) to 250-cell (blastocyst) stage. Embryos derived from embryo splitting have a lower chance of survival (for example, 50 percent versus 70 percent for intact embryos) when transferred to females. Survival of split embryos is further reduced after freezing, so splitting may not be an advantage in preservation. Embryos derived from the splitting of an original founder embryo may themselves be split to produce additional identical descendants, but the survival rate of the resulting embryos is markedly reduced.

Cloning by nuclear transplantation, originally developed for the mouse (McGrath and Solter, 1983), has produced viable embryos and offspring in cattle, pigs, rabbits, and sheep (First, 1991; Malter, 1992).

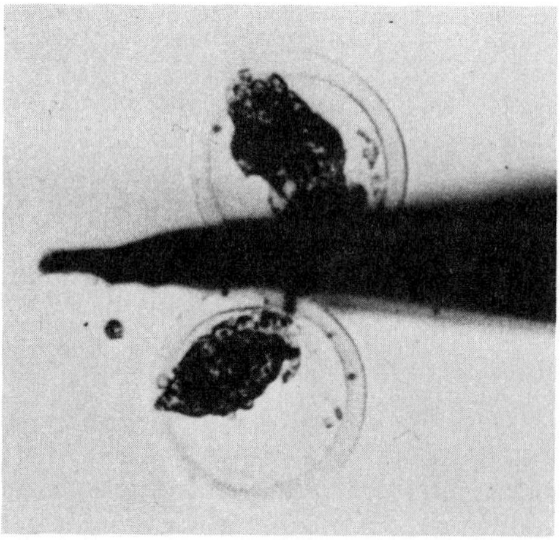

Using embryo transfer technology, a bovine embryo is split before implantation in a foster Boran mother to produce N'Dama-Boran crossbred calves. The first crossbred calves were produced in 1990; by the end of the year, 75 such calves were born. Large families of these cattle are needed for research at the International Laboratory for Research on Animal Diseases to locate the genes responsible for trypanotolerance, an ability of N'Dama and other indigenous livestock breeds in Africa to remain productive while infected with trypanosomes. Credit: International Laboratory for Research on Animal Diseases.

It is based on a technique for nuclear transfer first reported in the frog, *Rana pipiens* (Briggs and King, 1952). Nuclear transplantation involves the transfer of a nucleus or blastomere from a valuable embryo into a suitably receptive enucleated oocyte. After the reconstituted egg develops to a multicellular stage, it can be cloned again. Theoretically, large numbers of identical individuals could be produced using this method. Survival rates are still low, however, and the process has only been done through three generations.

For conservation it is important to retain as much genetic diversity as possible, which is accomplished by storing unrelated genotypes. Cloning does not increase the amount of genetic diversity; rather, it enables evaluation of stored samples without reducing the amount of diversity still preserved. It can also be used to provide a back-up collection or samples for multiple users. Clones would be of value, for example, in testing genetically identical individuals across different environments.

In any stock, live or cryopreserved, cloning would allow the immediate commercial exploitation of identified superior genotypes. Cloning has two drawbacks, however. First, only embryos have been cloned thus far, so duplicate (or multiple) frozen embryos must be retained until the clone has been evaluated for economic merit and then regenerated and multiplied. Second, the clone itself has no genetic variation and cannot be improved further. For continuous genetic improvement, the best clones would have to be mated to obtain new genetic combinations for another round of clonal selection.

Determining Sex

It is not yet possible to separate *X*-chromosome-bearing (female) and *Y*-chromosome-bearing (male) sperm and maintain normal sperm viability. Some progress has been made through flow cytometry, a method for rapidly measuring cells, based on the difference in the DNA content of *X* and *Y* sperm (Gledhill et al., 1982; Pinkel et al., 1985). Flow cytometric procedures can also be used for sorting sperm cells to produce samples enriched in *X* or *Y* sperm (Johnson et al., 1987). These sorted samples have been used successfully to produce live offspring (Johnson et al., 1989). The sex of embryos can be determined cytogenically from 3 or 4 cells taken at the 16- to 32-cell stage. More recently, male-specific antigen has been used to determine sex in embryos from the 16-cell to early blastocyst (100-cell) stages. DNA probes have been used to determine sex at all stages, and are available commercially in Australia, France, and the United States (Herr et al., 1990; Nibart et al., 1992).

Sexing of sperm or embryos, however, seems to be of little advantage in conservation. Storage costs are likely to be low relative to the costs of collection, so all sperm and embryos collected should be retained; determining the sex would only increase costs and losses in viability. If reasonable numbers of embryos are collected (for example, 100 to 200 to allow for losses), deviations from the 50:50 sex ratio are likely to be small. If very limited numbers of embryos are available, however, sexing may be used to decrease the chance of random genetic drift upon restoration of the population.

HEALTH STATUS OF GERMPLASM

One of the main limitations on the movement and use of genetic resources has been concern about the health status of the germplasm. The adverse effects of any imported diseases could negate potential gains from imported germplasm and deleteriously affect existing stocks. This concern has often barred exports and imports. Where transfer of germplasm has been allowed, stringent regulations and conditions have frequently been imposed. They usually involve expensive quarantine procedures over a period of time and a series of tests in both the exporting and importing countries for a specific set of diseases. Thus, the exchange of improved genetic stocks across countries has been expensive and often restricted in the past. Improved methods of disease detection and control and the use of frozen semen and embryos make international exchange of improved genetic stocks increasingly feasible. However, the potential to transmit disease agents must not be minimized or overlooked (see Appendixes B and C).

The preservation of rare stocks is more difficult. Limited numbers of individuals may be available as a source of germplasm, and they are unlikely to be kept under sanitary isolation conditions. Access to the animals may be limited in number and time, and their health status may be variable. The conditions under which the animals are kept and samples are collected may be marginal.

Another drawback is that the preserved stocks are retained for future use. Even if they satisfy today's health regulations, they may not satisfy those prevailing in the future. In this case all effort at conservation might be in vain. Alternatively, a new generation of the preserved stock would have to be produced and placed in quarantine until the new set of health conditions could be satisfied. Thus, it is important to separate animal germplasm from disease agents at the time of storage, if possible. One advantage of preserved germplasm would be that it will be free of any diseases that have developed or spread since its collection.

New technologies of semen screening (see Appendix C) and embryo washing (see Appendix B) hold considerable promise for improving the health status of germplasm. A method developed by Schultz et al. (1982) allows verification of the health status of semen samples. This technique may be valuable for testing semen from males known to be exposed to or infected with a disease. Semen immunoextension involves the treating of semen with specific antibodies to prevent disease transmission. Embryo washing and treatment remove or neutralize pathogens that may be found around the embryo in a layer called the zona pellucida. It has been effective against most pathogens tested to date (Singh, 1987). (See Appendix B.) These techniques have important implications for genetic conservation because they could allow collection of germplasm under ordinary farm conditions and the future testing of stored germplasm to meet new health standards.

RECOMMENDATIONS

Selection and crossbreeding have been used to make major contributions to animal improvement programs in many countries. Relatively little effort has been devoted, however, to the preservation of germplasm, which represents the biodiversity on which selection and genetic combinations are based.

Research on technologies that could benefit preservation and use of animal genetic resources should be continued and expanded.

New ideas, research results, and techniques are continuously being reported that can affect the methods and value of conservation. Research areas with the potential to enhance the preservation, management, and use of animal genetic resources include the following:

- Methods for collecting and cryopreserving sperm and embryos from minor breeds and under adverse field conditions.
- Reliable technologies to detect, eradicate, or control pathogenic agents that may be harbored in semen, ova, embryos, or live animals.
- Improved methods for cryopreserving the sperm and embryos of swine and avian species.
- Techniques for establishing and maintaining genetic libraries for each livestock species.
- Development of genomic maps for each livestock species.
- Population genetics studies of molecular methods to measure relationships within and among breeds so that priority breeds for preservation, use, or development can be identified.

- Expansion of the numbers of DNA markers (RFLPs and VNTRs) for livestock species.
- Techniques for marker-assisted selection in each major livestock species.

For selected species, DNA banks should be established to provide a potential source of specific genes for future livestock improvement.

A DNA store is a potential future source of specific genes. From genomic libraries particular genes of interest can be cloned. Specific genes from a DNA store may some day be of value for improving livestock through the use of directed gene introgression (breeding) or across species through transgenic technologies. A DNA bank could ensure sufficient genetic material for molecular characterization of the specific genes and DNA segments of a population, inexpensively stored, which could provide researchers with the material for molecular biological and gene mapping studies of that population. Although DNA stores may aid evaluation and use of germplasm, they should not be considered presently viable methods for conserving the genetic resources of livestock. Furthermore, the characterization of phenotypic attributes are not possible through DNA libraries.

Efforts should be made to link research on preservation, management, and use of livestock germplasm to analogous efforts in other fields of study.

Many studies on the human genome are being carried out. Knowledge gained from genome mapping studies of livestock can be of use in human mapping studies, just as information from emerging human maps can aid livestock efforts. Several zoos are working to develop methods of germplasm manipulation and storage for wild animals (Office of Technology Assessment, 1985, 1987) and their efforts benefit from livestock studies. Gains in molecular genetics derived from the study of plants also may have applications to animals.

5

National Programs

The more highly developed the animal industry of a nation, the more compelling is its need to make provisions for preserving unusual or unique populations and to ensure access to the genetic diversity that will be required for future breeding programs. Countries differ markedly in the size and character of their animal populations, in the requirements for managing animal genetic resources, and in their capability to initiate and support programs. Each country should determine which of its animal populations are endangered and which merit protection. Nevertheless, national programs to manage genetic resources will have many common technical and organizational elements. This chapter describes several national efforts and reviews the basic organizational elements of a national program.

EXAMPLES OF CURRENT NATIONAL EFFORTS

A number of national programs have been established, particularly in developed countries, and public and private initiatives are also in operation (Alderson, 1990a; Wiener, 1990). Well-known and successful examples are programs of the Nordic countries and the Rare Breeds Survival Trust in the United Kingdom (Alderson, 1990b).

In the United States, no organized federal program for maintaining the diversity of livestock animals exists. However, plans are under way for a national germplasm program, and the need for coordinated action at both national and international levels has been outlined (Council for Agricultural Science and Technology, 1984; Office of Technology Assessment, 1987). Private organizations actively involved in conservation efforts in the United States include the American Minor Breeds Conservancy, which has pioneered practical programs. Researchers affiliated with universities and agricultural experiment stations help to identify genetic resources or to maintain and develop germplasm resources, although not as much as breed associations or private industries do (Office of Technology Assessment, 1987).

Some developing countries, such as Argentina, Brazil, China, and India, have developed national strategies and established programs in the past 10 to 15 years. They are generally larger countries with many landraces, administrative and scientific infrastructures, trained personnel, and other resources. The following sections briefly summarize examples of public and private programs around the world.

Hungary

Preservation of old breeds started in Hungary in the 1950s as the result of individual efforts to save the last animals of traditional breeds. In 1973 the Hungarian Ministry of Agriculture and Food assigned responsibility for maintaining declining breeds to the Institute for Animal Breeding and Feed Control. The institute's program has focused on endangered, traditional Hungarian livestock populations, including Hungarian Grey Steppe cattle, Mangalica pigs, Racka sheep, and Curly Feathered (Sebastapol) geese. With financial support from the government, state farms and cooperative farms have been maintaining noncommercial herds. Cryogenic germplasm banks have also been established for current use and as long-term stores. The research program associated with the preserved herds and flocks has included studies on the evolution and genetic structure of the breeds and evaluations under present production conditions (Maijala et al., 1984).

Hortobagy National Park, located in east central Hungary, is one of the most effective examples of a national or state park established as a conservation area to preserve national or regional rare breeds (Henson, 1990). The conservation program at the park includes the protection of indigenous plants and wildlife as well as the perpetuation of regional skills and traditions of the plains farmers and their traditional breeds. The conservation program has its own state-funded budget, land at the park, and maintenance staff. Close links exist with university personnel, who control the breeding program and use the animals in extensive research. The park derives a portion of its budget from tourists.

Brazil

The Empresa Brasileira de Pesquisa Agropecuária (EMBRAPA, Brazilian Agricultural Research Enterprise) was one of the first research institutions in Latin America to establish a preservation program for endangered animal genetic resources, following the recommendations of the Food and Agricultural Organization (FAO) of the United Nations and the United Nations Environment Program (UNEP) (Primo, 1987). Through the Centro Nacional de Recursos Genéticos (CENARGEN, National Center for Genetic Resources) in Brazil, EMBRAPA is evaluating livestock populations of breeds that were originally introduced by European colonists. Over time these populations have acquired adaptive traits for Brazil's diverse ecological conditions. Today, however, many of them are in an advanced state of genetic dilution or in danger of extinction. Genetic conservation is achieved by maintaining breeding units or cryopreserved semen and embryos.

In addition to maintaining a germplasm bank, CENARGEN has undertaken characterization studies to describe the genetic attributes of cattle types. EMBRAPA is evaluating purebreds and crossbreeds in a number of environments and production systems. These activities are guided by a Brazilian national plan that was adopted in 1980 to implement activities for conserving animal genetic resources and to consider the conclusions of a technical consultation on conservation and management held in Rome by FAO and UNEP in 1980.

India

The animal genetic resources of the Indian subcontinent are large and diverse. Important livestock populations include draft cattle, milk buffaloes, carpet wool sheep, and highly prolific goats. Many Indian breeds are adapted to tropical heat and diseases and poor quality feeds. To study indigenous livestock, the Indian Council of Agricultural Research created the National Bureau of Animal Genetic Resources at Karnal in the late 1980s. It will characterize, evaluate, and catalog livestock genetic resources, establish a data bank and information service on these resources, and determine the need of and steps for conserving and managing them. In addition to nine national research centers and species institutes, the council has directorates on cattle and poultry improvement and breeding research projects that will cooperate with the bureau. The bureau will also undertake collaborative studies with state agricultural universities (Acharya, 1990).

Nongovernmental Projects

The Rare Breeds Survival Trust (RBST) in the United Kingdom is one of the best known of the nongovernment-funded organizations in the field of rare breed preservation. It is a privately funded and managed national charitable organization established in 1973 to ensure both the survival of rare farm breeds that are part of Britain's agricultural heritage and a pool of genetic variation for future livestock breeders (Alderson, 1985, 1986, 1990b). It has about 8,500 members, 1,000 of whom are active in owning and breeding small groups of animals of rare breeds. Support comes entirely from private contributions and corporate grants.

The RBST is an umbrella organization offering networking services to coordinate the voluntary activities of breeders of rare breeds,

Farm parks act as educational visitor centers while conserving rare breeds. Here visitors at the Cotswold Farm Park admire a Whitefaced Woodland sheep, an endangered breed in Great Britain that produces fine wool. This breed was developed by King George III in the 1790s by crossing local mountain sheep with Merino sheep. Credit: Philip Harvey, Cotswold Farm Park.

including farm parks. Farm parks are independent centers owned and operated by individuals, land owning charities, or city councils. They are recognized by the RBST in an approved scheme that identifies those farm parks genuinely involved in rare breed preservation. They have large collections of rare breeds in breeding units, and the parks are open to the public as educational and visitor centers. They have proved to be a successful means of encouraging public interest in and support of conservation work.

The conservation activities of the RBST include the establishment of procedures for defining rare breeds; the payment of subsidies to farmers for keeping rare cattle, horses, and pigs; recording pedigrees; and monitoring inbreeding, as needed, for rare breeds with no breed societies. Its staff members run workshops and livestock sales, give breeding advice, and produce a monthly magazine. The conservancy also seeks funding and helps to supply animals for scientific research and for evaluation of rare breeds. The only animals it owns are the North Ronaldsay sheep on the island of Linga Holm, Scotland.

In cooperation with the Milk Marketing Board for England and Wales, the RBST has established a cryogenic semen bank; samples from more than 100 bulls of 15 breeds are stored, and storage of samples from other breeds is planned. The RBST also collects semen for breeds that are not rare but are in a rapid state of change or contamination from other breeds.

The American Minor Breeds Conservancy (AMBC), located in Pittsboro, North Carolina, was formed in 1977 by a group of individuals who were interested in preserving rare breeds and stocks of livestock and poultry in North America. It works closely with similar groups, such as the Association of Living Historical Farms and Agricultural Museums, and zoo and wildlife conservation organizations. The AMBC has accomplished much work in identifying, rescuing, and restoring rare breeds and stocks, including Milking Devon and Dutch Belted cattle, Jacob and Navajo-Churro sheep, Dominique chickens, and Pilgrim geese. It has been developing recommended procedures for reproducing and maintaining rare poultry stocks.

The AMBC has about 2,200 members, 65 percent of whom keep livestock. Membership fees and special appeals support about half of the AMBC's work; the other half is financed by funds from benefactors and foundations (Crawford, 1990b).

ORGANIZATIONAL ELEMENTS OF A NATIONAL PROGRAM

Programs to manage livestock genetic resources should recognize the research goals already described, of which the sustaining of germplasm for ongoing improvement in commercial livestock populations is crucial. Plans to conserve genetic resources appropriate to this goal should be an integral part of livestock improvement programs. Similarly, the genetic characterization and evaluation for potential use should be, if possible, a prominent element of conservation programs. The dual goals of livestock improvement and conservation of genetic diversity must be addressed as dual components of an integrated genetic resource management program that aggressively pursues genetic improvement while protecting the genetic resources needed for improvement.

A national policy for conserving and using animal genetic resources is essential to building a program. A program must identify the risks to these resources and have priorities for guiding decisions about managing and allocating them. It should coordinate public and private efforts to preserve, evaluate, and use animal germplasm, and it should operate with related regional and global efforts.

A complete national program includes the following:

- Inventories and characterizations of animal populations, with special emphasis on indigenous types.
- Information recorded in an appropriate data base that can be accessed by potential users.
- Preservation of unique or endangered populations.
- Evaluation and international use of indigenous and exotic populations.
- Collaboration with other national and international programs.
- As appropriate, research in managing and using animal genetic resources.
- Strong links with genetic improvement programs to ensure they address conservation.

Documentation and Inventory

Characterization of all genetic resources, including their numbers and environments, and the development of comprehensive data bases are needed for all domestic animal species. Available data should include assessing the resource's risk of loss to provide a foundation for national and global decisions about conservation. However, characterization may require more time than is available for many populations now at risk. Therefore, the first priority must be preserving breeds or populations at risk to prevent loss of resources that may possess unique genes or gene combinations.

Knowledge of the kinds, numbers, status, structure, and rate of change of animal populations is essential to a national program. Information should be maintained in a data-base system that can be updated and supplemented. Advisory committees composed of scientists and other users can assist in determining conservation priorities.

Three kinds of information are important for setting management priorities. First, assessment must be made of vulnerability, including population size, structure, and status (increasing, decreasing, stable), degree of endangerment, and whether the population is at risk of being diluted or replaced by other populations. Second, phenotypic and genetic uniqueness should be determined. Finally, livestock populations should be characterized and evaluated to aid in selection for various economic, research, or breeding applications.

Small or declining populations are particularly important because they are most vulnerable to genetic loss. If accurate population counts are not known, estimation of the numbers of breeding individuals is an important first step in establishing the data base. Populations must then be inventoried periodically to assess their current status.

Animal Genetic Resources Data Base

Establishment of a national data base may require only part-time professional and technical support. Thus, several countries may pool efforts and professional expertise to support a regional data base, thereby making more efficient use of limited resources. Regional or international organizations that assist with planning and managing data bases require clear operating definitions of their roles and services relative to those of national programs. In addition, regional or global data banks need to be assured of consistent funding, either from cooperating nations or through international support.

Conservation of Unique and Endangered Populations

The maintenance of working herds, flocks, or populations is the predominant method for conserving animal genetic diversity in the commercial sector. However, for rare stocks, especially those that are not economically competitive, conservation through commercial sector activities may not be effective. Rare stocks may be better preserved through specially designed programs for live conservation and continued improvement, which also focus on indigenous types with recognized potential value. These programs may need financial support from public and private sources.

Cryogenic storage of gametes or embryos is another way of preserving rare stocks. Samples for cryopreservation should be documented as fully as possible. Guidelines have been developed by the FAO for characterizing health status, environmental conditions at the time of collection, population statistics for production traits, animal and tissue sampling procedures, individual animal records, and details on the treatment of samples (for example, embryo washing protocols during processing for freezing).

The genetic diversity within major production stocks also should be regularly monitored. Declines in genetic variation and narrowing of the genetic base as a result of intensive selection or inbreeding should be assessed. Conservation of the germplasm of primary production stocks should not be overlooked.

Evaluation and International Use of Indigenous and Exotic Populations

Production and marketing systems around the globe vary greatly. National animal improvement programs must consider the sustainable capacities of specific production systems and use appropriate genetic stocks. The selection of stocks used for breeding will also be influenced by the local environment, which can significantly affect expression of a particular genotype and, thus, affect productivity.

Access to novel germplasm in the form of improved stocks can be a major benefit to animal breeders. However, the contributions of adapted indigenous populations also can be valuable, especially when there is an adverse production environment.

The introduction of inappropriate genetic stocks can have long-term and deleterious consequences. Important adapted traits can be lost from local breeds, and indigenous stocks may be totally replaced by imported breeds. Indigenous landraces and populations can be the sole resources for environmentally adaptive traits or particular production characters. Thus when indigenous populations are threatened by imported stocks, national actions are necessary to ensure their conservation.

Collaboration with Other Programs

National programs should interact with other national, regional, and global efforts to conserve and manage animal genetic resources. International cooperation should include participation in regional or global data bases and standardization of criteria for the safe movement of animal germplasm. Information on the status and availability of specific animal germplasm could assist the exchange and use of germplasm among countries.

Some developed countries will have the ability, desire, and re-

A foreman at the Corriedale ranch in New Zealand shows a prized Corriedale ram. This breed was developed in 1882 through private sector efforts to combine wool and meat production traits in one breed. The Corriedale is a Merino-Lincoln crossbreed. Credit: James P. Blair, ©1992 National Geographic Society.

sources to participate in multilateral or bilateral efforts to aid the maintenance of genetic diversity in developing countries. International aid could support preserving indigenous breeds or locally adapted populations that are threatened by replacement or serious dilution and lack adequate evaluation. Support could also come through the sponsorship of training programs in areas such as quantitative genetics, use of breed resources, conservation and breeding to heighten awareness for knowledgeable and responsible sustainable development of animal agriculture using imported and indigenous breeds.

Great strides have been made in recent years in procedures and technologies to control diseases. Methods for testing animals are very sensitive, and infected and disease-free individuals can be identified readily. The quarantine of live animals can, however, be prohibitively expensive. Import and export regulations are not always based on the latest scientific evidence. An important development in terms of the international movement of germplasm is the emergence of embryo washing techniques (see Appendix B). In light of new technology, it will be essential that nations coordinate policies and required protocols to ensure the safe, reliable exchange of animal germplasm, without undue restrictions.

Opportunities for Research

Research is an important element of national germplasm management programs. Many aspects of research related to livestock improvement, such as reproductive physiology, cryopreservation, and molecular biology, are being conducted in universities, international centers, and private industry. A national program should determine its need to pursue new research efforts, which should be of practical and immediate application, and in areas not addressed by other programs. Research areas needing attention include: characterization and evaluation of indigenous populations; cryopreservation technology for avian semen, swine and poultry embryos, and semen and embryos of minor species and wild relatives; collection of oocytes, in vitro fertilization, embryo splitting, nuclear transfers, and cloning of embryos; development of effective field techniques for collecting semen and embryos; molecular techniques to measure genetic differences within and between breeds; development and utilization of DNA (deoxyribonucleic acid) libraries; techniques for DNA transfer; and methods for fertility and reproductive enhancement.

GERMPLASM CONSERVATION IN DEVELOPED AND DEVELOPING COUNTRIES

In developed countries the numbers of breeds, varieties, and strains of most of the agriculturally important animal species are declining. Trends toward industrialized animal production systems in many developed countries have led to demands from industries and consumers for greater uniformity in animal performance and animal products. Although the decline of many breeds is generally recognized, there is some disagreement on the relationship between the loss of breeds and the narrowing of the genetic base. The situation varies among species. For example, replacement of poultry stocks with those adapted to controlled, closed environments occurs faster than with cattle stocks that must still be adapted to varying local conditions and seasons.

In many developing countries, where the environment, disease, climate, and the feed supply limit breeding efforts, the most important characteristic of livestock is still their ability to thrive in harsh circumstances. The stresses of climate, feed and water, altitude, disease, and pests and parasites, combined with farmer preferences, have imposed powerful selection forces in the tropics and subtropics. Breeding programs are needed to improve productivity without decreasing adaptation to the environment. In many tropical areas, temperate improved breeds are being imported, while local ones are neglected. This policy has worked well with dairy cattle in many subtropical and dry tropical areas, leading to the wider distribution of Holstein cattle. It has generally failed, however, in the humid tropics. Furthermore, success can be maintained only if intensive production inputs, such as feed supply and medication for disease control, are maintained.

In harsher environments, local breeds can often be best improved through carefully planned crossbreeding and selection techniques (Mason, 1984). Bali cattle, for example, are smaller than European cattle and are low milk producers, but they exhibit high fertility, minimal fat deposition, efficient water utilization, and an ability to thrive in hot humid climates and on poor quality feed. Their size, agility, and ease of training also make them particularly suitable for work in small fields of irregular shape. In many parts of Latin America, Africa, and southern Asia, the goat, which can live and thrive on poor ranges, is the principal source of meat and milk for the subsistence farmer. Selection programs to improve milk yield and meat production in tropical goat breeds are needed.

In developing countries, efforts in recent decades to increase animal productivity nearly always involved the introduction of imported germplasm. The technique of artificial insemination and the capability to freeze semen, which can then be shipped easily and stored indefinitely, have encouraged these attempts at change. The newly developing techniques of embryo transfer offer another means for introducing germplasm, but success has been mixed, in part because of a lack of appropriate infrastructures for artificial insemination, embryo transfer, and communication and education among small livestock producers. Only when conditions permit the whole transfer of modern production technology and genetic stocks has it been possible to replace existing populations with introduced ones. Replacement has been possible more frequently with poultry or pigs than with ruminants (J. Hodges, consultant, Mittersill, Austria, March 1992). Increasingly, however, local indigenous breeds are being diluted or replaced. Much of the unique germplasm of livestock in developing countries may thus disappear.

Some developing countries have recently established programs to conserve animal genetic resources (Hodges, 1990a; Wiener, 1990). These are generally large countries with many indigenous breeds, a well-developed administrative and scientific infrastructure, trained personnel, and laboratory resources. For example, Argentina, Brazil, China, and India have national strategies involving the use of live animals and cryogenic techniques to preserve endangered breeds. Activities also exist in several smaller developing countries such as Botswana, Morocco, and Zimbabwe (Setshwaelo, 1990). These and other national efforts have been aided by the FAO and the UNEP (Hodges, 1992a), which have developed methodologies, provided training courses and publications, and held expert meetings focusing on the needs and opportunities for improved conservation and management. However, smaller developing countries are often unable to devote resources to develop their own programs.

RECOMMENDATIONS

The primary aim of national animal germplasm conservation programs is to ensure the maintenance and accessibility of sufficient livestock genetic diversity to support increases in animal production in a variety of systems, and to accommodate changes in selection goals, production environments, and market requirements. Conservation includes maintenance, preservation, and propagation of endangered breeds that may be unique or possess potentially useful heritable qualities. To be effective, national programs should also facilitate the characterization, evaluation, and use of animal germplasm. International programs are needed to coordinate and support national efforts and to provide support for information management and training in germplasm conservation and use (see Chapter 6).

National programs for preserving, managing, and using livestock genetic resources must be established in developed and developing countries and suited to different needs.

The principal work of conserving animal genetic resources must be planned at the national level. Public and private activities should be integral parts of the national plan. Given the substantial differences in needs and appropriate approaches in individual countries, the organization of national conservation programs will vary widely among countries. No single best strategy or program exists, although some elements are common to all. For example, a national program must document national populations, with emphasis on indigenous breeds. Further, it should participate in maintaining a data base on indigenous and imported genetic resources. This task may be accomplished most efficiently in cooperation with similar regional and global efforts. For example, governments could cooperate on a regional basis to share responsibilities, resources, facilities, and technical expertise.

An essential first step for successful animal genetic resources management at the national level is identification of a center, office, or individual charged with the responsibility of monitoring and, to the extent possible, fostering conservation of national genetic resources. This person or office could cooperate with international efforts and other national programs, and should become a center of knowledge about the status of national livestock resources.

Mechanisms for providing bilateral and multilateral support to enable developing nations to preserve, manage, and use their livestock genetic resources must be developed.

Multilateral and bilateral aid programs will be required in many cases to help establish national livestock improvement and conservation efforts in developing countries. Technical expertise from the developed countries should also be readily available. Conservation of diversity is a global concern, and given the greater threat to uncharacterized, indigenous livestock populations in developing countries, the support of improvement and conservation programs in developed nations is essential. Ultimately, livestock production and improvement programs of both developed and developing countries will benefit from international support and assistance efforts.

Operational programs that implement strategic genetic conservation activities and prevent further loss of potentially valuable livestock genetic resources should be established.

The issues surrounding the conservation and use of animal genetic resources have been discussed and debated in numerous forums for more than 30 years. The general conclusions from these meetings are similar: diversity in global livestock populations is potentially threatened, valid reasons exist to conserve diversity in livestock populations, and international coordination of strategic conservation activities is highly desirable. Further delineation of the need for conservation and management of animal genetic resources is not required.

Information and Data

Increased efforts are needed to inventory and characterize unique and endangered populations and breeds, particularly in developing countries.

Effective conservation and management of genetic resources in livestock will require basic information on animal populations: population sizes, structures, and trends, phenotypic characteristics, productivity, and environmental conditions under which they have been raised. Much descriptive and analytic information has been gathered on livestock breeds, particularly in developed countries. However, in developing countries, far less has been accomplished in terms of organized studies to inventory, characterize, and compare livestock populations, and often there is still insufficient information on which to base management decisions.

6

International Programs and a Global Mechanism

Changing patterns of livestock production, with the goal of improving animal performance through genetic selection and improved production methods, are increasingly placing the global genetic diversity of livestock at risk. Since the 1950s, when freezing the semen of many species became possible, artificial insemination has become widely used. The international movement of livestock germplasm has also grown with the aim of increasing livestock production in developing countries (Hodges, 1991). In the developed world, the displacement of traditional breeds is well advanced.

Few adaptation problems were encountered when traditional temperate breeds were replaced by other temperate breeds. In the tropics, however, problems in adaptation have become more evident. Livestock improvement programs have needed to focus on the appropriate mix of exotic and indigenous genes for a given environment, the genetic effects of crossbreeding, and methods for achieving the desired gene mix.

Although the changing patterns in livestock breeding have been recognized for more than 40 years, effective action to prevent losses of potentially valuable germplasm has been limited. Funds for germplasm conservation have been difficult to justify for several reasons. The actual extent and rate of genetic depletion are not readily quantifiable, and, until recently, there has been little consensus on either the most effective approaches to conserving and managing endangered populations or to identifying priority candidates for conservation efforts. In addition, it is difficult to assess current and future values for conserved material in relation to the expense of conservation. The last decade, however, has seen some progress toward a resolution of these difficulties including the development of new scientific methods for ex situ conservation. International awareness of the impact of human activities on biological diversity has been raised and there is a growing realization that something must be done (Food and Agriculture Organization, 1992; Hodges, 1991, 1992b).

This chapter provides an overview of major international activities, particularly during the past 10 years. It reviews information sources and data bases, proposed programs to conserve livestock genetic resources, conferences and studies that have identified concerns and assessed needs, regional programs to store germplasm, pilot projects, alternative methodologies, and key related issues of plant genetic resource programs. It identifies the need for a global mechanism for conserving livestock germplasm and examines the potential for developing global leadership.

INTERNATIONAL PROGRAMS

Concern over the potential loss of valuable genetic diversity for livestock has increased in recent years (Henson, 1992), although the issue was recognized decades ago. For more than 40 years, international activity with animal genetic resources has been focused primarily through the Food and Agriculture Organization (FAO) of the United Nations. In 1948, FAO published its first study on the subject (Food and Agriculture Organization, 1948), and over the next 25 years, studies or projects have led to many different publications on livestock species. These reports largely describe the breeds in the environments in which they occur, production gains through improved health care and nutrition, and genetic improvements gained largely by selection within the indigenous breed, often with slow and limited success.

Although FAO has taken a lead position, many other international and nongovernmental organizations are actively involved in livestock germplasm issues. Studies or other activities have been undertaken by international agricultural research centers, especially the International Livestock Center for Africa (ILCA), International Laboratory for Research on Animal Diseases (ILRAD), and Centro Agronómico Tropical de Investigación y Enseñanza, (Tropical Agriculture Research and Training Center). Other concerned and active organizations include regional institutions such as the Society for the Advancement of Breeding Researches in Asia and Oceania (SABRAO), European Association for Animal Production (EAAP), Asociación

Latinoamericana de Producción Animal (ALPA, Latin American Association of Animal Production), Danubian Countries Alliance for Gene Conservation in Animal Species, Inter-African Bureau for Animal Resources (IBAR) of the Organization of African Unity (OAU), and SABRAO as well as Arabic and Nordic countries. The United Nations Environment Program (UNEP), United Nations Educational, Scientific, and Cultural Organization, World Conservation Union (IUCN), and World Resources Institute, also have indicated their interest.

Various conferences and studies have addressed managing and conserving livestock genetic resources and have contributed to an increasing awareness of the issues involved (Henson, 1992; Hodges, 1991). Early references to animal conservation were made at a 1959 symposium on germplasm (Hodgson, 1961) and at a 1964 poultry conference (European Poultry Conference, 1964). The need for concerted conservation action for livestock genetic resources has been discussed in a variety of forums. A continuing global forum on genetic resources has been organized since 1974 under the International Committee for World Congresses on Genetics Applied to Livestock Production.

Livestock genetic resources are also on the agenda of the United Nations Development Program and the World Bank, and they were included within biological diversity at the United Nations Conference on Environment and Development (UNCED) in Brazil in June 1992. The issue is scheduled as a major topic at the 1993 World Conference on Animal Production, to be held in Canada. In addition, an international nongovernmental organization, Rare Breeds International (RBI), was created in 1991 in Budapest, Hungary, with private and governmental conservation organizations from 30 countries as members. Although FAO is primarily interested in conservation efforts in the developing world, it has expressed interest in coordinating efforts with RBI for developed countries.

Efforts on a Global Level

The growing international awareness of a deteriorating global environment in the early 1970s led to a United Nations conference on the environment and to the creation of the UNEP (Hodges, 1991). The FAO was one of the first specialized agencies of the United Nations to join with the newly established UNEP in formulating projects that addressed the need for development and conservation. Thus, in 1973, FAO started a pilot program with UNEP to conserve animal genetic resources (Office of Technology Assessment, 1987). Preliminary inventories of endangered livestock populations indicated the need for further action, but it soon became clear that activity on a single country basis was inadequate. Many issues and needs were common to all developing countries. The FAO publications of the period show interest moving to regional activities and species (for example, dairy cattle breeding in the humid tropics).

In 1980, the FAO and UNEP held the Technical Consultation on the Conservation and Management of Animal Genetic Resources to identify priority areas and activities on a global and regional basis (Hodges, 1991). At that time, the term *management* included development activities; thus the concepts of improvement and conservation were already linked. All member governments of FAO and UNEP were invited to participate in the meeting. It produced a comprehensive series of recommendations covering the reasons for genetic erosion in animal resources; the rationale for linking conservation with improved management; establishment of regional activities, data banks, and germplasm banks; promotion of appropriate research in areas likely to aid conservation and reproduction; training of nationals from developing countries; study of health barriers to germplasm movement; breeding programs for conservation and improvement; special studies of little-known animal populations in the People's Republic of China and the former Soviet Union; and publication of a newsletter.

Another result was the creation in the 1980s of an FAO and UNEP Joint Expert Advisory Panel on Animal Genetic Resources Conservation and Management, which consisted of 36 scientists, representing various areas of the globe, whose expertise covered the principal disciplines within animal breeding and genetics and the major species of domestic animals. The panel has presented technical papers and recommendations in the areas of conservation by management, animal genetic resources data banks, cryogenic storage of germplasm, genetic engineering, training programs (such as embryo transfer methods), principles for improving indigenous animals in the tropics, principles for preserving endangered species and breeds in the tropics, and education and training in the tropics (Food and Agriculture Organization, 1984a,b; Hodges, 1987).

Activities related to training were also jointly sponsored by the FAO and UNEP (Hodges, 1991). A course on livestock genetic resources conservation and management was hosted by the University of Veterinary Science in Budapest, Hungary, in 1983 for 18 scientists from developing countries, and a training manual was published (Bodó and Hodges, 1984; Bodó et al., 1984). A manual on establishing and operating animal germplasm banks was prepared following a 1989 workshop held in Hanover, Germany (Food and Agriculture Organization, 1990).

The FAO and UNEP have worked with regional organizations to arrange meetings on animal genetic resources. In 1976 the OAU recommended that the IBAR establish an expert committee on animal genetic resources in Africa. With support from the FAO and UNEP, the committee was formed. A joint meeting was held with FAO, UNEP, and IBAR in 1983 to identify high potential and endangered livestock in Africa. In 1978 a conference was held in Colombia in cooperation with ALPA, and ALPA thereafter established a commission on evaluating and conserving animal genetic resources in Latin America. In Asia, SABRAO established an expert committee on animal genetic resources in 1977.

In 1980 the EAAP, in response to FAO's recommendations for action at national and regional levels, established the Working Party on Animal Genetic Resources (European Association for Animal Production, 1984). It serves as a coordinating body for European activities. For example, the Nordic countries (Denmark, Finland, Iceland, Norway, and Sweden) have cooperative efforts under way to document and preserve animal genetic resources.

The activities defined by the FAO and UNEP technical consultation in 1980, and monitored by the Joint Expert Advisory Panel, were completed between 1982 and 1989. Methodology studies and field trials in many countries were completed, documented, and published for characterizing breeds; operating animal genetic data banks; sampling endangered populations; collecting semen, embryos, and DNA; conducting health checks and monitoring; and establishing cryogenic germplasm.

In 1989, the FAO Committee on Agriculture (COAG) reviewed the programs on animal genetic resources (Hodges, 1991). Subsequently, the FAO council affirmed the recommendations of the COAG and called for the program to be expanded and further developed. As a result, the FAO reviewed its animal genetic resources program and convened an expert consultation in September 1989 to provide guidance on global program directions (Wiener, 1990). The papers and recommendations covered the following: institutional, legal, and financial aspects of a global program; biotechnology in preserving animal germplasm; live animal preservation; status of endangered breeds and establishment of a world watch list and early warning system; and technical and organizational aspects. Unfortunately, due to a poor economic climate and the absence of an organizational system for a worldwide program, these recommendations did not lead to a global program at that time. The FAO continued with some activities supported by the regular program and UNEP.

The FAO Program on Managing Animal Genetic Resources

In January 1992 FAO announced the initiation of a program to conserve and develop the livestock and poultry genetic resources of developing nations (Cunningham, 1992). The program's five elements addressed conservation of important or valuable germplasm resources and assistance in developing national programs for livestock and poultry improvement. The elements encompassed initiatives in a global inventory, breed preservation, indigenous breed development and conservation, gene technologies, and the development of an international legal framework. An Expert Consultation in April 1992 examined the proposed program and provided further guidance and refinement of the plan, including definition of the operating mechanism for a program (Food and Agriculture Organization, 1992; Hodges, 1992b). The ILCA is assisting with the implementation of this program in sub-Saharan Africa, in collaboration with FAO (Rege and Lipner, 1992).

Global Inventory

Building from the EAAP data base developed at the Hanover School of Veterinary Medicine, the program will develop a data base of breed descriptions and population statistics for livestock and poultry species that will, ultimately, contain a comprehensive global inventory. The program will also develop a World Watch List of Threatened Livestock (Cunningham, 1992; Food and Agriculture Organization, 1992; Hodges, 1992b). Classification of breeds will be in accordance with a scheme published by FAO (Food and Agriculture Organization, 1992; Henson, 1992; Hodges, 1992b).

Breed Preservation

Endangered breeds or populations will be evaluated and those that are also unique will be the objects of country-based conservation strategies. Such strategies could include ex situ (cryopreservation of semen and embryos) and in situ (breeding herds or flocks) methods. Where appropriate, FAO will work to establish regional banks for ex situ materials. The intent is to link conservation and use efforts in developing countries.

Indigenous Breed Development and Conservation

The intent is the genetic improvement of selected indigenous breeds of livestock and poultry. The original plan of FAO was to focus on 12 unique populations, chosen on the basis of regional importance and genetic uniqueness (Cunningham, 1992). That list was expanded by the Expert Consultation to include six cattle populations, three buffalo populations, five each of sheep and goat populations, four pig populations, three camelids, two horses, and eight poultry forms (Table 6-1).

TABLE 6-1 Breeds, Strains, or Populations of Livestock and Poultry Suggested as Having Highest Priority for Conservation by Management

	Geographic Region[a]		
Species	Asia	Africa	Latin America
Cattle	Sahiwal (1)	N'Dama (2) Kenana (4) Boran (5)	Criollo (3) Guzerá (6)
Buffalo	Murrah (1) Nili-Ravi (3) Swamp (2)		
Sheep	Javanese Thin-tailed (3) Awassi (5)	Djallonké (1) D'man (4)	Pelibüey (2)
Goats	Damascus (2) Jamnapari (3)	Fouta Djallon (1) Boer (5)	Moxotó (4)
Pig	Taihu (1) Min (4)	West African (3)	Piau (2)
Camelidae			Huacaya alpaca (1) Suri alpaca (2) Qara llama (3)
Horses	Akhal-Teke (1)		Pantaneiro (2)
Poultry	Domestic duck (4) Muscovy duck (3a) Chicken (5a)	Guinea fowl (2) Chicken (5b)	Indigenous turkey (1) Muscovy duck (3b) Chicken (5c)

NOTE: Conservation by management is defined as that aspect of conservation by which a sample, or the whole of an animal population, is subjected to planned genetic change with the aim of sustaining, utilizing, restoring, or enhancing the quality or quantity of the animal genetic resource and its products of food, fiber, or draught animal power.

[a]Numbers in parentheses indicate priority order within species.

SOURCE: Adapted from Food and Agriculture Organization. 1992. Expert Consultation on the Management of Global Animal Genetic Resources, Rome, Italy, April 7–10, 1992. Rome, Italy: Food and Agriculture Organization of the United Nations.

Gene Technologies

To promote development of DNA technologies, gene mapping, genetic engineering methods, and ex situ storage methods (e.g., cryopreservation), FAO will seek to establish a research fund from extrabudgetary sources for support of relevant projects. The intent is to foster the development of more efficient and cost-effective methods of conservation and use. The Expert Consultation urged FAO to develop opportunities to train developing country scientists in these emerging technologies (Food and Agriculture Organization, 1992; Hodges, 1992b).

An International Legal Framework

With global distribution and use of germplasm comes the need for clearly established understandings about ownership, exchange, and compensation. FAO, in consideration of existing instruments and proposals (such as the Convention on Biological Diversity signed in June 1992 at UNCED), will seek relevant agreements to ensure an appropriate legal framework for global animal genetic resource activities (Food and Agriculture Organization, 1992; Hodges, 1992b). In addition, the Expert Consultation recommended that FAO promote and encourage the formation of national and appropriate regional programs for the conservation and management of livestock and poultry (Food and Agriculture Organization, 1992; Hodges, 1992b).

INFORMATION SOURCES AND DATA BASES

The information available on livestock populations generally takes three forms: published reports, analyses from qualified individuals, and unpublished production records. Publications include research papers in periodicals, reports, proceedings, theses, and books. Most publications are catalogued and available through libraries, while some printed materials, such as extension service bulletins, may not be indexed and may be difficult to identify and obtain.

The publications in the series entitled *FAO Animal Production and Health* are a valuable contribution to the characterization and management of global animal genetic resources. They include proceedings of meetings of the Expert Advisory Panel on Animal Genetic Resources Conservation and Management (Food and Agriculture Organization, 1984a,b; Hodges, 1987), descriptions of specialized populations (for example, Awassi sheep [Epstein, 1985]; Przewalski horse [Food and Agriculture Organization, 1986a]), and descriptions of the livestock populations for an entire country or region (for example, livestock of China [Cheng, 1985]; livestock of the former Soviet Union [Dmitriev and Ernst, 1989]). In addition, the FAO publishes, at irregular intervals, *Animal Genetic Resources Information,* a bulletin containing short articles, news items, and book reviews about general topics.

Data Bases

The need for data bases on animal genetic resources has been recognized in Asia (Society for the Advancement of Breeding Researches in Asia and Oceania, 1980), Latin America (Müller-Haye, 1981), Africa (Organization of African Unity et al., 1985), and Europe (European Association for Animal Production, 1984). Probably the most important source of documentation for animal genetic resources is the data base of the Commonwealth Bureau of Animal Breeding and Genetics (operating under the Commonwealth Agricultural Bureaux [CAB]) (Turton, 1984). It consists of the *Animal Breeding Abstracts,* a journal that has been in existence for more than 55 years, and a computer-stored version of the journal, *CAB Abstracts,* which was started in 1973 and is maintained by the Knight Ridder Company's Dialog Information Services, Miami, Florida; the European Space Agency Retrieval Service, Frascati, Italy; and Deutsches Institut für Medizinische Dokumentation und Information (DIMDI, German Institute for Medical Documentation and Information), Köln, Germany. As an abstracting service, the CAB data base does not fulfill all content and retrieval requirements for use in animal genetic resources. Another, more serious problem is that it contains only a small amount of information from developing countries (Hodges, 1986).

The FAO and the UNEP have implemented pilot trials to establish data bases for animal genetic resources in Africa, Asia, and Latin America. Methodologies for data handling, including data entry and retrieval, and descriptor lists for the major species have been developed. During data collection, it was found that only 25 to 30 percent of important information for the characterization of animal populations was included in the CAB data base (Food and Agriculture Organization, 1986b,c,d). Much of the information from developing countries is not readily available to an abstracting service because it is not published or has a restricted distribution. Further, the need for foreign language translations inevitably reduces the volume of material that can be covered. Thus, an urgent need exists for establishing comprehensive data bases to store and make available more complete information on the genetic resources of animal populations.

Workshops were held in the three regions to evaluate the pilot trials, and conclusions were published in three reports (Food and

Agriculture Organization, 1986b,c,d). The joint FAO and UNEP trials provide valuable information about the advantages and disadvantages of various alternatives for data collection and processing in developing countries and have identified problems to be resolved. The reports discuss alternative computer programs for animal genetic resource data bases; establishment of regional versus global data bases; recommendations on, as well as cost of, software and hardware for computerized data bases; and descriptor lists for cattle, buffalo, pigs, sheep, and poultry. The descriptor lists provide for environmental as well as genetic characterizations and consist of an open-ended system for including unlimited data.

The EAAP Working Party on Animal Genetic Resources has conducted extensive surveys to identify the number and degree of endangerment of breeds of cattle, horses, sheep, pigs, and goats (Maijala et al., 1984, 1985). In 1985 the EAAP decided to establish, at the Hanover School of Veterinary Medicine in Germany, a central data base for European countries. In 1988 an agreement was reached with the FAO to cooperate with the school to create the EAAP and FAO Animal Genetic Data Bank (Simon, 1990). The data bank is based on the use of simplified descriptors identified by the FAO and UNEP and contains information on about 700 entries. Each entry refers to a single breed or breed variety within a country. The two principal information categories are (1) census data on populations of breeds and (2) genetic characterization data. Because experience is still limited, the data base in Hanover remains the principal international animal genetic resources data base that is routinely operational. Unfortunately it has received little information from developing countries apart from India and China, which have staff who were trained at the Hanover data bank.

The ILCA has gained much data-base experience. In particular, the ILCA has been engaged in documenting unpublished material that can be used for describing populations and in analyzing field and experiment station records, the results of which can be used for data bases (Brumby and Trail, 1986). Starting in 1976, a documentation staff from ILCA has microfilmed more than 10,000 documents in 17 African countries. Their efforts constitute a valuable source of information for describing populations that can eventually be used for an animal genetic resources data base. Titles, descriptors, and abstracts of documents have been stored in a computerized data base and are accessible on-line (Trail, 1984). The ILCA has also established a service that assists in data analysis, if it cannot be done locally, and contributes to producing information for entry in data bases.

Denmark, Finland, Iceland, Norway, and Sweden have set up the

Nordic Gene-Bank for maintaining information on local populations of sheep, goats, horses, cattle, and pigs. Plans call for including data on fur-bearing farm animals, poultry, rabbits, reindeer, and bees.

Since the mid-1980s, information on the water buffalo has been collected and recorded in a data base at the International Buffalo Information Center at Kasetsart University in Bangkok, Thailand. The International Development Research Center, Ottawa, Canada, is funding the project.

To begin to remedy the lack of information from developing countries, FAO set up a data bank in 1991 that includes all areas of the world except Europe, but including the former Soviet Union. It has been designed to combine the animal genetic information from the large variety of FAO livestock reports, publications, and projects. The FAO data base uses software developed at the Hanover data bank and the *World Dictionary of Livestock Breeds* as a framework (Mason, 1988). The dictionary is the only orderly world list of domestic breeds. It provides breed names and synonyms, indicates where breeds are found, and gives a basic description of the origin, physical appearance, and main uses of each breed. Although the dictionary has little characterization data, it offers a unique classification of most breeds. The FAO data bank has names of 1,974 breeds (buffalo, cattle, goat, horse, pig, sheep) in Africa, Asia, Latin America, North America, Oceania, and the former Soviet Union, and allows one entry per breed, even if the breed is found in more than one country (Ruane, 1992). Of these entries, 30 percent have some information on production, 26 percent have some information on population size, and 18 percent have information on both. Within 5 years, FAO, as part of its new initiative on global animal genetic resources, plans to finish this first global inventory of animal genetic resources.

A joint effort is being made by the FAO and UNEP to describe breeds based on surveys of non-European countries, with good responses from Asia and Latin America to date (International Livestock Center for Africa, 1992). Some developing countries do not have data on certain parameters, such as average milk yield or total feed consumed. However, numerous workshops and seminars have produced a considerable volume of information about the description of populations, which will help in establishing data bases (Müller-Haye and Gelman, 1981; Organization of African Unity et al., 1985).

CREATING REGIONAL STORES OF FROZEN GERMPLASM

Considerable logistic and technical difficulties, such as obtaining technical expertise for collection, maintaining constant supplies of liquid nitrogen, or establishing ownership of frozen samples, are involved in creating repositories of frozen germplasm, particularly in developing countries. There is growing concern, however, that if action is not taken soon to preserve the genetic material of endangered breeds, many of these breeds will be lost (Alderson, 1990a; Wiener, 1990).

A recent activity of the FAO and UNEP has been the development of plans for the cryogenic storage of sperm and embryos from rare and endangered breeds of livestock in regional animal germplasm banks (Food and Agriculture Organization, 1990; Hodges, 1990b). The material to be deposited in the germplasm banks would be for long-term storage as insurance against unexpected future changes, rather than for current use.

The necessary infrastructures for these regional cryogenic germplasm banks were negotiated with governments in seven developing countries. Lack of funding has so far restricted the inflow of germplasm from other countries in the regions. Nevertheless, the infrastructure is in place, and the FAO and the UNEP are working with national governments to train personnel and identify samples meriting deposition (Hodges, 1990b). Training courses for nationals were held in Brazil and in Asia in 1991 and 1992, respectively. Most of the participants were from countries that did not have national germplasm banks. Two germplasm banks are located in Asia (China and India), two in Africa (Ethiopia and Senegal), and three in Latin America (Argentina, Brazil, and Mexico). Because of regional differences in the status of foot-and-mouth disease in Latin America, the Mexican germplasm bank will be a subregional bank. For example, samples from Central America, if stored in South America, may not be eligible for return to the countries of origin.

The regional animal germplasm banks will store the blood, gametes, and embryos of donor animals, enabling future development of gene libraries. Blood samples may also be of value for diagnostic testing. Plans call for linking the regional animal germplasm banks with the FAO global data bank.

RELATED ISSUES OF PLANT GENETIC RESOURCE PROGRAMS

National and international programs for conserving livestock genetic resources have not had the widespread support that exists for crops. The issues bearing on collection, creation of germplasm banks, and exchange of resources differ in many respects between animals and plants. However, important lessons are to be learned from plant conservation activities, which embrace a wide range of genetic diversity, from wild relatives to highly selected lines. Ownership of plant genetic resources is often not well defined, and issues surrounding it have caused considerable international debate. The species of primary concern for livestock conservation are essentially all the domesticates, and ownership has been clearly established. Proprietary rights, however, are likely to become a more controversial issue with the advent of laws allowing patents on genetically engineered animals.

In addition, as nations come to recognize they may hold unique, potentially valuable breed types, they may wish to impose limitations on the sale, exchange, or movement of these breeds. They may also wish to retain control over semen or embryos deposited in regional or global germplasm banks and to impose restrictions on the export of wild relatives, should an interest develop in the commercial potential of the genetic material of those species.

National and international activities in plant genetic resources can provide guidance for emerging efforts to conserve animal genetic resources. Several national plant genetic resource conservation programs have been developed in the past 30 to 40 years. In 1974 the International Board for Plant Genetic Resources (IBPGR) was established within the FAO by the Consultative Group on International Agricultural Research (CGIAR) and given a mandate to conserve plant genetic resources. The IBPGR coordinates international efforts in plant conservation and has fostered the development of new efforts. It oversees a network of base collections of important species located within various national programs and at international agricultural research centers. Pending approval of a headquarters agreement by the Italian government, IBPGR will soon become the International Plant Genetic Resources Institute (IPGRI)—a CGIAR center independent of FAO.

GLOBAL CONSERVATION OF ANIMAL GERMPLASM

Given the potential for genetic losses in livestock populations, the often limited knowledge of the level of genetic diversity present in populations, and the benefits of collaboration among nations, the need to establish a global mechanism for conserving animal germplasm becomes compelling. Global leadership is needed to

- Inventory and describe livestock populations;
- Establish a uniform data base for collecting and disseminating data concerning animal populations, institutions, research, and germplasm banks;

• Provide technical guidance to nations in structuring their national animal germplasm conservation strategies;

• Develop training programs in animal germplasm conservation; and

• Assist in obtaining financial support for establishing national or regional programs.

Currently, the FAO, with cooperation from the UNEP, is the only international agency addressing animal germplasm conservation on a global scale. The FAO and UNEP activities have greatly contributed to an understanding of the need for conserving genetic diversity in domestic animals, but, in the past, they have fallen short of providing effective programs or leadership to the international community.

Criteria for a Global Program

The task of promoting, coordinating, and instituting global management of animal genetic resources is significant. Leadership will require an organization to be technically capable and have the international position to bring about real progress. The committee identified several criteria for evaluating the suitability of particular options for global leadership.

Donor Confidence

The institution must be able to provide or secure sufficient funding to accomplish program goals. This capability requires the somewhat intangible quality that can be termed "donor confidence"—based on the institution's record of success, the donor can be reasonably assured of the institution's continued success.

Scientific Capability

The institution must also be able to provide the scientific expertise essential to developing and guiding a program. It is not essential that the institution have research capacity, since essential research can be arranged with appropriate scientists or institutions on a cooperative basis. Rather, the capacity must exist for the institution to make the scientific and technical aspects of genetic resources conservation, management, and use a central concern in program development. The program also should be responsible to a board or committee designed to provide technical and scientific oversight and recommendations about program activities.

Administrative Expertise

Within the institution there must be an understanding of the essential processes that guide international organizations. This understanding will be crucial to the development and execution of international agreements and cooperative activities. There must also be the understanding and expertise to forge appropriate legal and international agreements with regard to exchange, ownership, and use of livestock genetic resources.

Consultative Mechanism

As this report has demonstrated, there are many interests and differing views concerning managing animal genetic resources. A global program must enfold this diversity of views and activities. The program must have a mechanism for including the perspectives of national programs, university researchers, nongovernmental organizations, industry, and others.

Options for a Global Program

Three potential options for developing the necessary global leadership were identified and considered by the committee. Each takes advantage of institutional infrastructures already in place. They are discussed below.

An Institute Within the Consultative Group on International Agricultural Research

The CGIAR was established in 1972 to enable private, bilateral, and multilateral organizations to collaborate in addressing the problems of agricultural development. The relative freedom and flexibility of organization, management, funding, and oversight shared by the CGIAR centers are unique among international bodies. The centers have a strong emphasis on basic food crops, several maintain important collections of plant germplasm, and one—the IBPGR—promotes and coordinates activities related to plant genetic resources. Two centers, the ILCA in Ethiopia and ILRAD in Kenya, have programs devoted to animals.

A model similar to IBPGR could be used to incorporate animal genetic resources into the CGIAR. The ILCA and ILRAD undertake animal genetic research, but neither has global responsibility for animal germplasm conservation. The ILRAD's genetic research is primarily at the molecular level, including contributions to mapping the bovine genome. The ILCA's research involves characterization of indigenous cattle, sheep, and goat populations in sub-Saharan Africa. However, these institutions could provide a base for other genetic resources activities. Additional national programs or institutions would need to be identified for collaboration in other regions.

A proposal to create a center for animal genetic resources within the CGIAR is not new (Council for Agricultural Science and Technology, 1984; Office of Technology Assessment, 1987). Such a program could build on the experience and reputation of the CGIAR and IBPGR. The center would coordinate national and regional efforts to conserve and manage livestock genetic resources. It could foster appropriate research, provide support for information and germplasm exchange, and assume a leadership role.

As an alternative to a new institution, the activities of the new IPGRI might be expanded to include animals. This would require a significant change in the mandate of IPGRI and a probable restructuring of its advisory boards. The new activities would, however, benefit from the history of experience in crop genetic resources.

Significant new financial resources would be required to organize a new program within the CGIAR or to expand the existing IBPGR/IPGRI. The CGIAR is cautious about undertaking new activities that would increase competition for resources among the centers (Consultative Group on International Agricultural Research, 1990), and its donors generally prefer to emphasize research rather than service functions. Recent decisions will expand CGIAR's activities to include agroforestry, aquaculture, and irrigation, but several other initiatives were deferred or turned down (Consultative Group on International Agricultural Research, 1990). The addition of work in animal genetic resources could entail considerable study before action is taken. Donors who support the IBPGR might be reluctant to expand its mandate to include animals without assurance of increased financial resources.

Expanded Efforts of the Food and Agriculture Organization

The FAO could assume leadership for developing strategies and programs for animal germplasm conservation. Currently the Animal Production and Health Division of the FAO, working with the UNEP, is engaged in the only global initiative in animal germplasm conservation. In addition, the FAO collaborates with regional programs, including ALPA, EAAP, ILCA, and SABRAO. It has information and experience related to animal germplasm issues in different regions.

The FAO, however, has not yet had adequate funding to fully develop many of the actions recommended by its advisory panels.

The FAO has been concerned with conserving animal genetic resources for more than 2 decades. It has a standing expert panel on conservation and management of animal genetic resources and program officers within the Animal Production and Health Division with designated responsibility for conservation. In addition, the FAO is accustomed to working with national governments, and it has an experienced bureaucracy in place that could manage a conservation program. Thus, there is institutional experience to support the FAO in undertaking this initiative.

The FAO's history of involvement with conservation issues and programs carries some costs, however. There has been some conflict of views about several issues related to plant genetic resources as they affect developed and developing nations. It is possible that disagreements, if they continue, could carry over to animal conservation programs and hamper their effectiveness. Furthermore, given the important role for the private sector, the FAO's limited involvement with private organizations could pose a difficulty.

Expanded Support of an Existing Conservation Organization

Most conservation-related organizations are not structured to deal with agricultural germplasm conservation. Some reorientation of programs, funding, philosophy, and purpose might be required. In general, these organizations are skilled in international conservation activities and have highly dedicated staffs.

An organization that might undertake responsibility for animal genetic resources is the World Conservation Union, formerly the International Union for Conservation of Nature and Natural Resources, with headquarters in Switzerland. The IUCN has been involved in conservation activities for more than 40 years. Its membership includes governments, government agencies, and national and international nongovernmental organizations. It has experience in developing data bases for supporting conservation activities, monitoring natural areas, and fostering international cooperation for conservation (Office of Technology Assessment, 1987).

Expansion of the IUCN's mandate would be required to include livestock genetic resources. However, philosophical conflicts between agricultural necessities and the organization's environmental priorities are likely. Efforts developed through conservation organizations such as the IUCN would need the expertise and experience of groups such as the FAO to fully address agricultural livestock conservation, management, and use.

The most feasible strategy may well involve a combination of these three options, with each institution taking responsibility for that aspect most clearly within its mandate, experience, and capabilities. However, a continuing call for global leadership would remain.

RECOMMENDATIONS

The committee examined alternatives for establishing a global mechanism for managing livestock genetic resources. To be effective, an organization must have a clear mandate, competent leadership, consistent funding, the confidence of its members and clients, and the freedom and flexibility to act expeditiously. The program's costs, effectiveness, and continuity must be properly established and monitored.

National authorities must ultimately define the policies and strategies appropriate for their nation's needs, but the support and counsel of an international scientific body are highly desirable. National and regional programs can function in the absence of global coordination, but many actions common to all national programs can be performed more effectively if coordinated at the international level.

A global mechanism should be established to provide leadership and support to nations to ensure the adequate conservation of livestock genetic resources.

An international organization would provide leadership to coordinate and facilitate national and regional efforts. It can enhance national activities, foster cooperation, and assemble experienced specialists to develop priorities and guidelines for global cooperation. Global leadership could be achieved through the creation of a new organization or through modification of existing organizations, as previously discussed. However, none of the three options presented fully meets all criteria outlined. No one existing organization has the elements necessary to fully undertake the tasks. A global strategy will likely involve the cooperation of several institutions that would take responsibility for the aspects of conservation most appropriate to their experiences and capabilities. Thus it becomes essential that leadership for an international program rests with an institution that has the capacity to organize and coordinate these diverse interests.

In the past it has been difficult for FAO to move beyond the initial phases of program description and planning. However, new confidence and encouragement has come with the development of the current FAO program. Plans are moving forward for active cooperation with international centers and scientific experts to accomplish definite goals. With the identification of particular populations to receive attention, FAO has the potential to address a specific and achievable task. Further, the scientific expertise present in the leadership of the FAO program provides a significant measure of confidence that donors will see real results. Finally, moving forward with a program at FAO would allow for efforts to begin relatively rapidly. As noted above, execution of a global program through a new or expanded CGIAR center or through a nongovernmental conservation organization would require significant restructuring or institutional development.

Considering the urgency to begin a program of global leadership in the coordination, promotion, and institution of animal genetic resources activities, FAO was considered by the committee to be an appropriate institution for these efforts. Further, because FAO has been active in this arena for several decades, establishment of a program would require minimal development of new institutional mechanisms. It is expected, however, that in implementing such a program, the FAO will call on the expertise of other national and international institutions to achieve its goals.

Comprehensive data bases are needed to store and make available information on animal genetic resources.

Data bases should be set up to assemble information that describes animal populations according to size, geographic distribution, objectives, production system, and priority for conservation. They should also contain quantitative information on important traits and estimates of phenotypic, genetic, and environmental parameters, together with information on the environmental conditions under which the data were obtained. They can be established at national or international levels. Some efforts are already under way (Hodges, 1990b; Simon, 1990). The FAO Global Data Bank for Domestic Livestock, modeled on the Animal Genetic Data Bank in Hanover, Germany, should provide much-needed leadership and guidance for national and regional data bases.

References

Acharya, R. M. 1990. India's effort in conservation and management of indigenous livestock and poultry genetic resources and creation of data base. Pp. 175–190 in Animal Genetic Resources: Strategies for Improved Use and Conservation, J. Hodges, ed. Rome, Italy: Food and Agriculture Organization of the United Nations.

Alderson, L. 1985. The conservation of animal genetic resources in Great Britain. Anim. Genet. Resour. Inf. 4:26–31.

Alderson, L. 1986. Mobilization of the forces of society for the conservation of animal genetic resources. Anim. Genet. Resour. Inf. 5:1–5.

Alderson, L., ed. 1990a. Genetic Conservation of Domestic Livestock. Wallingford, U.K.: CAB International.

Alderson, L. 1990b. The work of the Rare Breeds Survival Trust. Pp. 32–34 in Genetic Conservation of Domestic Livestock, L. Alderson, ed. Wallingford, U.K.: CAB International.

Andersson, L. R. 1986. Genomic hybridization of bovine class II major histocompatibility genes. I. Extensive polymorphism of DO alpha and DO beta genes. Anim. Genet. 17:95–112.

Baccus, R., N. Ryman, M. H. Smith, D. Reuterwall, and D. Cameron. 1983. Genetic variability and differentiation of large grazing mammals. J. Mammal. 64:109–120.

Baker, C. M. A., and C. Manwell. 1991. Population genetics, molecular markers and gene conservation of bovine breeds. Pp. 221–304 in Cattle Genetic Resources, C. G. Hickman, ed. Amsterdam: Elsevier.

Beckmann, J. S., and M. Soller. 1983. RFLP's in genetic improvement: Methodologies, mapping and costs. Theor. Appl. Genet. 67:35–43.

Beckmann, J. S., Y. Kashi, E. M. Hallerman, A. Nave, and M. Soller. 1986. Restriction fragment length polymorphisms among Israeli Holstein-Friesen dairy bulls. Anim. Genet. 17:25–38.

Benyo, D. F., G. K. Haibel, H. B. Laufman, and J. L. Pate. 1991. Expression of major histocompatibility complex antigens on the bovine corpus luteum during the estrous cycle, luteolysis, and early pregnancy. Biol. Reprod. 45(2):229–234.

Bindon, B. M., and L. R. Piper. 1986. Booroola (*F*) gene: Major gene affecting ovarian function. Pp. 67–93 in Genetic Engineering of Animals—An Agricultural Perspective, W. J. Evans and A. Hollaender, eds. New York: Plenum.

Bodó, I., and J. Hodges. 1984. Manual for Training Courses on the Animal Genetic Resources Conservation and Management, Vol. 2. Budapest, Hungary: University of Veterinary Science.

Bodó, I., V. Buvanendran, and J. Hodges. 1984. Manual for Training Courses on the Animal Genetic Resources Conservation and Management, Vol. 1. Budapest, Hungary: University of Veterinary Science.

Briggs, R., and T. J. King. 1952. Transplantation of living nuclei from blastula cells into enucleated frogs' eggs. Proc. Natl. Acad. Sci. USA 38:455–463.

Brumbaugh, J. A. 1989. Gene transfer in poultry. Pp. 15–16 in The Nebraska Poultry Report, M. M. Beck, ed. Lincoln: Cooperative Extension Service, University of Nebraska.

Brumby, P. J., and J. C. M. Trail. 1986. Animal breeding and productivity studies in Africa. Int. Livestock Center Afr. Bull. 23:23–27.

Cahaner, A., and P. B. Siegel. 1986. Evaluation of industry breeding programs for meat-type chickens and turkeys. Pp. 337–346 in Proceedings, 3rd World Congress on Genetics Applied to Livestock Production, Vol. 10, G. E. Dickerson and R. K. Johnson, eds. Lincoln: University of Nebraska.

Capecchi, M. R. 1989. Altering the genome by homologous recombination. Science 244:1288–1292.

Chambers, S. M., and J. W. Bayless. 1983. Systematics, conservation and the measurement of genetic diversity. Pp. 349–363 in Genetics and Conservation: A Reference for Managing Wild Animal and Plant Populations, C. M. Schonewald-Cox, S. M. Chambers, B. MacBryde, and L. Thomas, eds. Menlo Park, Calif.: Benjamin/Cummings.

Chambers, J. R., J. S. Gavora, and A. Fortin. 1981. Genetic changes in meat type chickens in the last twenty years. Can. J. Anim. Sci. 61:555–563.

Cheng, P. L. 1985. Livestock Breeds of China. FAO Animal Production and Health Paper No. 46. Rome, Italy: Food and Agriculture Organization of the United Nations.

Christensen, A., D. A. Sorensen, T. Vesteryaard, and P. van Kemenade. 1986. The Danish pig breeding program: Current system and future developments. Pp. 143–148 in Proceedings, 3rd World Congress on Genetics Applied to Livestock Production, Vol. 10, G. E. Dickerson and R. K. Johnson, eds. Lincoln: University of Nebraska.

Clutton-Brock, J. 1981. Domesticated Animals from Early Times. Austin: University of Texas Press.

Consultative Group on International Agricultural Research (CGIAR). 1990. International Centers Week 1990: Summary of Proceedings and Decisions. Washington, D.C.: CGIAR Secretariat, World Bank.

Cotswold Farm Park. 1985. Rare Breeds Survival Centre, Guiting Power. Norwich, U.K.: Jarrold & Sons.

Council for Agricultural Science and Technology. 1984. Animal Germplasm Preservation and Utilization in Agriculture. Report No. 101. Ames, Iowa: Council for Agricultural Science and Technology.

Crawford, R. D. 1990a. Experience with *in situ* preservation of poultry breeds. Pp. 142–150 in Animal Genetic Resources Conservation. A global programme for Sustainable Development. FAO Animal Production and Health Paper No. 80. Rome, Italy: Food and Agriculture Organization of the United Nations.

Crawford, R. D. 1990b. The work of the American Minor Breeds Conservancy, with special reference to the conservation of poultry breeds. Pp. 3–17 in Genetic Conservation of Domestic Livestock, L. Alderson, ed. Wallingford, U.K.: CAB International.

Cundiff, L. V., K. E. Gregory, R. M. Koch, and G. E. Dickerson. 1986. Genetic diversity among cattle breeds and its use to increase beef production efficiency in a temperate environment. Pp. 353–358 in Proceedings, 3rd World Congress on Genetics Applied to Livestock Production, Vol. 9, G. E. Dickerson and R. K. Johnson, eds. Lincoln: University of Nebraska.

Cunningham, E. P. 1992. Conservation and development of animal genetic resources: FAO outline programme. Pp. 49–53 in Management of Global Animal Genetic Resources, J. Hodges, ed. Rome, Italy: Food and Agriculture Organization of the United Nations.

Cunningham, E. P., and O. Syrstad. 1987. Crossbreeding *Bos indicus* and *Bos taurus* for Milk Production in the Tropics. FAO Animal Production and Health Paper No. 68. Rome, Italy: Food and Agriculture Organization of the United Nations.

Dmitriev, N. G., and L. K. Ernst, eds. 1989. Animal Genetic Resources of the USSR. FAO Animal Production and Health Paper No. 65. Rome, Italy: Food and Agriculture Organization of the United Nations.

Enfield, F. D. 1988. New sources of variation. Pp. 215–218 in Proceedings of the Second International Conference on Quantitative Genetics, B. S. Wier, E. J. Eisen, M. M. Goodman, and G. Namkoong, eds. Sunderland, Mass.: Sinauer Associates.

Epstein, H. 1969. Domestic Animals of China. Slough, U.K.: CAB Books.

Epstein, H. 1971. The Origin of the Domestic Animals of Africa, Vol. 1. New York: African Publishing.

Epstein, H. 1985. The Awassi Sheep, with Special Reference to the Improved Dairy Type. FAO Animal Production and Health Paper No. 57. Rome, Italy: Food and Agriculture Organization of the United Nations.

European Association for Animal Production (EAAP). 1984. Conservation of animal genetic resources in Europe. Final report of an EAAP working party. Livestock Prod. Sci. 11:3–22.

European Poultry Conference. 1964. Proceedings, Second European World Poultry Science Association Congress. Bologna, Italy: Accademia Nazionale di Agricultura.

First, N. L. 1991. New advances in reproductive biology of gametes and embryos. Pp. 1–21 in Animal Applications of Research in Mammalian Development, R. A. Pedersen, A. McLaren, and N. L. First, eds. Plainview, N.Y.: Cold Spring Harbor Laboratory Press.

Food and Agriculture Organization (FAO) of the United Nations. 1948. Breeding Livestock Adapted to Unfavourable Environments. Rome, Italy: Food and Agriculture Organization of the United Nations.

FAO. 1984a. Animal Genetic Resources Conservation by Management, Data Banks and Training. FAO Animal Production and Health Paper No. 44/1. Rome, Italy: Food and Agriculture Organization of the United Nations.

FAO. 1984b. Animal Genetic Resources: Cryogenic Storage of Germplasm and Molecular Engineering. FAO Animal Production and Health Paper No. 44/2. Rome, Italy: Food and Agriculture Organization of the United Nations.

FAO. 1986a. The Przewalski Horse and Restoration to Its Natural Habitat in Mongolia. FAO Animal Production and Health Paper No. 61. Rome, Italy: Food and Agriculture Organization of the United Nations.

FAO. 1986b. Animal Genetic Resources Data Banks. 1. Computer Systems Study for Regional Data Banks. FAO Animal Production and Health Paper No. 59/1. Rome, Italy: Food and Agriculture Organization of the United Nations.

FAO. 1986c. Animal Genetic Resources Data Banks. 2. Descriptor Lists for Cattle, Buffalo, Pigs, Sheep and Goats. FAO Animal Production and Health Paper No. 59/2. Rome, Italy: Food and Agriculture Organization of the United Nations.

FAO. 1986d. Animal Genetic Resources Data Banks. 3. Descriptor Lists for Poultry. FAO Animal Production and Health Paper No. 59/3. Rome, Italy: Food and Agriculture Organization of the United Nations.

FAO. 1990. Manual on Establishment and Operation of Animal Gene Banks. Food and Agriculture Organization of the United Nations, Rome, Italy. Photocopy.

FAO. 1992. Expert Consultation on the Management of Global Animal Genetic Resources, Rome, Italy, April 7–10, 1992. Rome, Italy: Food and Agriculture Organization of the United Nations.

Foreman, M. E., and J. E. Womack. 1989. Genetic and synteny mapping of parathyroid hormone and beta hemoglobin in cattle. Biochem. Genet. 21:541–550.

Freeman, B. M., and N. Bumstead. 1987. Transgenic poultry: Theory and practice. Worlds Poult. Sci. J. 43(3):180–189.

Fries, R., R. Hediger, and G. Stranzinger. 1988. The loci for parathyroid hormone and β-globin are closely linked and map to chromosome 15 in cattle. Genomics 3:302–307.

Fries, R., J. S. Beckmann, M. Georges, M. Soller, and J. E. Womack. 1989. The bovine gene map. Anim. Genet. 20:3–29.

Gannon, F., R. Powell, T. Barry, T. G. McEvoy, and J. M. Sreenan. 1990. Transgenic farm animals. J. Biotechnol. 16(3/4):155–170.

Gibson, J. P., and C. Smith. 1989. The incorporation of biotechnologies into animal breeding strategies. Pp. 203–231 in Animal Biotechnology: Comprehensive Biotechnology, First Supplement, M. Moo-Young, L. A. Babiuk, and J. P. Phillips, eds. Elmsford, N.Y.: Pergamon.

Gledhill, B. L., D. Pinkel, D. L. Garner, and M. A. Van Dilla. 1982. Identifying *X*- and *Y*-chromosome-bearing sperm by DNA content: Retrospective perspectives and prospective options. Pp. 177–191 in Prospects for Sexing Mammalian Sperm, R. P. Amann and G. E. Seidel, Jr., eds. Boulder: Colorado Associated University Press.

Glodek, P. 1982. Experimental results with pigs. Pp. 243–253 in Proceedings of the Second World Congress on Genetics Applied to Animal Production, Vol. 6. Madrid, Spain: Graficas Orbe.

Goddard, M. E. 1991. The global black and white cattle population. J. Dairy Sci. 74(Suppl. 1):233.

Gonzalez, P., M. J. Tunon, and M. Vallejo. 1987. Genetic relationships between seven Spanish native breeds of cattle. Anim. Genet. 18:249–256.

Graham, E. F., M. L. Schmehl, and R. C. M. Deyo. 1984. Cryopreservation and fertility of fish, poultry and mammalian spermatozoa. Pp. 4–23 in Proceedings of the 10th Annual Conference on Artificial Insemination and Reproduction. Columbia, Mo.: National Association of Animal Breeders.

Hallerman, E. M., J. L. Theilmann, J. S. Beckmann, M. Soller, and J. E. Womack. 1988. Mapping the bovine prolactin and rhodopsin genes in hybrid somatic cells. Anim. Genet. 19:123–131.

Harris, H. 1966. Enzyme polymorphisms in man. Proc. Roy. Soc. Lond. Ser. B 164:298–310.

Hawkins, D. R. 1988. The basic resource—The national cow herd. Northeast region. Pp. 48–54 in Proceedings of the 1988 National Beef Cattle Conference. Stillwater: Oklahoma State University.

Hayashi, S., K. Kobayashi, J. Mizuno, K. Saitoh, and S. Hirano. 1989. Birth of piglets from frozen embryos. Vet. Rec. 125:43–44.

Henson, E. L. 1985. North American Livestock Census. (Updated May 1988.) Pittsboro, N.C.: American Minor Breeds Conservancy.

Henson, E. L. 1990. The organization of live animal preservation programmes. Pp. 103–117 in Animal Genetic Resources: A Global Programme for Sustainable Development, G. Wiener, ed. FAO Animal Production and Health Paper No. 80. Rome, Italy: Food and Agriculture Organization of the United Nations.

Henson, E. L. 1992. In Situ Conservation of Livestock and Poultry. FAO Animal Production and Health Paper No. 99. Rome, Italy: Food and Agriculture Organization of the United Nations.

Herr, C. M., K. I. Matthaei, U. Pertzak, and K. C. Reed. 1990. A rapid *Y*- chromosome detecting ovine embryo sexing assay. Theriogenology 33(1):245.

Hill, W. G. 1982. Prediction of response to artificial selection from new mutations. Genet. Res. 40:255–278.

Hill, W. G., and P. D. Keightley. 1988. Interrelations of mutation, population size, artificial and natural selection. Pp. 57–70 in Proceedings of the Second International Conference on Quantitative Genetics, B. S. Wier, E. J. Eisen, M. M. Goodman, and G. Namkoong, eds. Sunderland, Mass.: Sinauer Associates.

Hines, H. C., C. A. Kiddy, E. W. Brum, and C. W. Arave. 1969. Linkage among cattle, blood and milk polymorphisms. Genetics 62:401–412.

Hodges, J. 1986. Animal genetic resources in the developing world: Goals, strategies, management and current status. Pp. 474–485 in Proceedings, 3rd World Congress on Genetics Applied to Livestock Production, Vol. 12, G. E. Dickerson and R. K. Johnson, eds. Lincoln: University of Nebraska.

Hodges, J., ed. 1987. Animal Genetic Resources: Strategies for Improved Use and Conservation. FAO Animal Production and Health Paper No. 66. Rome, Italy: Food and Agriculture Organization of the United Nations.

Hodges, J. 1990a. Animal Genetic Resources: A Decade of Progress, 1980–1990. Rome, Italy: Food and Agriculture Organization of the United Nations.

Hodges, J. 1990b. Review of regional animal gene banks and recommendations from Hannover Workshop on associated topics raised by the Tenth Committee on Agriculture. Pp. 51–58 in Animal Genetic Resources: A Global Programme for Sustainable Development, G. Wiener, ed. FAO Production and Health Paper No. 80. Rome, Italy: Food and Agriculture Organization of the United Nations.

Hodges, J. 1991. Management of Global Animal Genetic Resources: Review of Past and Present Activities and Potentials for the Future. Rome, Italy: Food and Agriculture Organization of the United Nations.

Hodges, J. 1992a. Conservation of rare breeds. In Proceedings of the Second International Conference on Animal Genetic Resources. Wallingford, U.K.: CAB International.

Hodges, J, ed. 1992b. Management of Global Animal Genetic Resources. FAO Animal Production and Health Paper No. 104. Rome, Italy: Food and Agriculture Organization of the United Nations.

Hodgson, R. E., ed. 1961. Germ Plasm Resources Symposium, Chicago. Publication No. 66. Washington, D.C.: American Association for the Advancement of Science.

Human Gene Mapping 10. 1989. Proceedings of the Tenth International Workshop on Human Gene Mapping, New Haven, Conn. Cytogenet. Cell Genet. 51.

International Laboratory for Research on Animal Diseases (ILRAD). 1991. ILRAD 1990: Annual Report of the International Laboratory for Research on Animal Diseases. Nairobi, Kenya: International Laboratory for Research on Animal Diseases.

International Livestock Center for Africa. 1992. African Animal Genetic Resources: Their Characterization, Conservation, and Utilization. Addis Ababa, Ethiopia: International Livestock Center for Africa.

Jeffreys, A. J. 1979. DNA sequence variants in the G-gamma, A-gamma, delta and beta-globin genes of man. Cell 18:1–10.

Johnson, L. A., J. P. Flook, M. V. Look, and D. Pinkel. 1987. Flow sorting of *X*- and *Y*-chromosome-bearing spermatozoa into two populations. Gamete Res. 16:1–9.

Johnson, L. A., J. P. Flook, and H. W. Hawk. 1989. Sex preselection in rabbits: Live births from *X* and *Y* sperm separated by DNA and cell sorting. Biol. Reprod. 41:199–203.

Kidd, K. K., W. H. Stone, C. Crimella, C. Carenzi, M. Casati, and G. Rognoni. 1980. Immunogenetic and population genetic analysis of Iberian cattle. J. Anim. Blood Groups Biochem. Genet. 11:21–38.

Lalley, P. A., S. J. O'Brien, N. Creau-Goldberg, M. T. Davisson, T. H. Roderick, G. Echard, J. E. Womack, J. M. Graves, D. P. Doolittle, and J. N. Guidi. 1987. Report of the committee on comparative mapping, HGM9. Cytogenet. Cell Genet. 47:367–389.

Larsen, B. 1977. On linkage relations of ceruloplasmin and polymorphism (*Cp*) in cattle. J. Anim. Blood Groups Biochem. Genet. 8:111–113.

Leibo, S. P. 1986. Cryobiology: Preservation of mammalian embryos. Pp. 251–272 in Genetic Engineering of Animals, J. W. Evans and A. Hollaender, eds. New York: Plenum.

LeQuarré, A. S., R. Hanset, and G. Vassart. 1987. Genetic variation of the bovine thyroglobulin gene studied at the DNA level. Anim. Genet. 18:41–50.

Lewontin, R. C., and J. L. Hubby. 1966. A molecular approach to the study of genic heterozygosity in natural populations. II. Amount of variation and degree of heterozygosity in natural populations of *Drosophila pseudoobscura*. Genetics 54:595–609.

Maguire, J. E., R. Ehrlich, W. I. Fuels, and D. S. Singer. 1990. Regulation of expression of a class I major histocompatibility complex transgene. J. Reprod. Fertil. 41:59–62.

Maijala, K., A. V. Cherekaev, J.-M. Devillard, Z. Reklewski, G. Rognoni, D. L. Simon, and D. E. Steane. 1984. Conservation of animal genetic resources in Europe. Final report of an EAAP (European Association for Animal Production) working party. Livestock Prod. Sci. 11:3–22.

Maijala, K., D. L. Simon, and D. E. Steane. 1985. Report by the Working Party on Animal Genetic Resources. Paper presented at the 36th Annual Meeting of the European Association for Animal Production, Kallithea, Halkidiki, Greece, September 30–October 3, 1985.

Malter, H. E. 1992. Micromanipulation in animal husbandry: Gene alteration and cloning. Pp. 47–83 in Micromanipulation of Human Gametes and Embryos, J. Cohen, H. E. Malter, B. E. Talansky, and J. Grifo, eds. New York: Raven.

Mason, I. L., ed. 1984. Evolution of Domesticated Animals. New York: Longman.

Mason, I. L. 1988. World Dictionary of Livestock Breeds. Wallingford, U.K.: CAB International.

McGrath, J., and D. Solter. 1983. Nuclear transplantation in the mouse embryo by microsurgery and cell fusion. Science 220:1300.

Müller-Haye, B. 1981. Posibilidades de programas de acción internacionales para la conservación y evaluación de recursos genéticos animales en América Latina con referencia especial al ganado Criollo. Pp. 141–156 in Recursos Genéticos Animales en América Latina, B. Müller-Haye and J. Gelman, eds. FAO Animal Production and Health Paper No. 22. Rome, Italy: Food and Agriculture Organization of the United Nations.

Müller-Haye, B., and J. Gelman, eds. 1981. Recurso Genéticos Animales en América Latina. FAO Animal Production and Health Paper No. 22. Rome, Italy: Food and Agriculture Organization of the United Nations.

Nakamura, Y., M. Leppert, P. O'Connell, R. Wolff, T. Holm, M. Culver, C. Martin, E. Fujimoto, M. Hoff, E. Kumlin, and R. White. 1987. Variable number of tandem repeat (VNTR) markers for human gene mapping. Science 235:1616–1622.

National Research Council (NRC). In press. Managing Global Genetic Resources: Agricultural Crop Issues and Policies. Washington, D.C.: National Academy Press.

NRC. 1991. Microlivestock: Little-Known Small Animals with a Promising Economic Future. Washington, D.C.: National Academy Press.

Nei, M. 1987a. Genetic distance and molecular phylogeny. Pp. 193–223 in Population Genetics and Fishery Management, N. Ryman and F. Utter, eds. Seattle: University of Washington Press.

Nei, M. 1987b. Molecular Evolutionary Genetics. New York: Columbia University Press.

Nei, M., and F. Tajima. 1981. DNA polymorphism detectable by restriction endonucleases. Genetics 97:145–163.

Nevo, E. 1978. Genetic variation in natural populations: Patterns and theory. Theor. Population Biol. 13:121–177.

Nibart, M., G. Kohen, L. Esposito, P. Baudu, J. M. Thuard, P. Desmettre, and P. Thibier. 1992. Rapid bovine sexing by DNA probe: Field results. Paper presented at the 12th International Congress on Animal Reproduction, The Hague, August.

O'Brien, S. J., W. G. Nash, D. E. Wildt, M. E. Bush, and R. E. Benveniste. 1985. A molecular solution to the riddle of the giant panda's phylogeny. Nature 317:140–144.

Office of Technology Assessment (OTA), U.S. Congress. 1985. Grassroots Conservation of Biological Diversity in the United States. OTA-BP-F-38. Washington, D.C.: U.S. Government Printing Office.

OTA. 1987. Technologies to Maintain Biological Diversity. OTA-F-330. Washington, D.C.: U.S. Government Printing Office.

Organization of African Unity (OAU), Food and Agriculture Organization of the United Nations, and United Nations Environment Program. 1985. Animal Genetic Resources in Africa: High Potential and Endangered Livestock. Proceedings of the Second OAU Expert Committee Meeting on Animal Genetic Resources in Africa. Nairobi, Kenya: Eleza Services.

Pinkel, D., D. L. Garner, B. L. Gledhill, S. Lake, D. Stephenson, and L. A. Johnson. 1985. Flow cytometric determinations of the proportion of *X*- and *Y*-chromosome-bearing sperm in samples of purportedly separated bull sperm. J. Anim. Sci. 60:1303–1307.

Pond, W. G., R. A. Merkel, L. D. McGilliard, and V. J. Rhodes. 1980. Animal Agriculture: Research to Meet Human Needs in the 21st Century. Boulder, Colo.: Westview.

Primo, A. T. 1987. Conservation of animal genetic resources: Brazil National Programme. Pp. 165–173 in Animal Genetic Resources: Strategies for Improved Use and Conservation, J. Hodges, ed. Rome, Italy: Food and Agriculture Organization of the United Nations.

Pursel, V. G., C. A. Pinkert, K. F. Miller, D. J. Bolt, R. G. Campbell, R. D. Palmiter, R. L. Brinster, and R. E. Hammer. 1989. Genetic engineering of livestock. Science 244:1281–1288.

Rall, W. F., and G. M. Fahy. 1985a. Vitrification: A new approach to embryo cryopreservation. Theriogenology 23:220.

Rall, W. F., and G. M. Fahy. 1985b. Ice-free cryopreservation of mouse embryos at –196°C by vitrification. Nature (Lond.) 313:573–575.

Rege, J. E. O., and M. E. Lipner, eds. 1992. African Animal Genetic Resources: Their Characterization, Conservation, and Utilisation. Addis Ababa, Ethiopia: International Livestock Center for Africa.

Rendel, J. 1958. Studies of cattle blood groups. II. Parentage tests. Acta Agric. Scand. 8:131–161.

Rodero, A., E. Camacho, E. Rodero, I. Serrano, and J. V. Delgano. 1990. Rare native breeds of Andalusia: Census, characterization and conservation strategy. Pp. 59–64 in Genetic Conservation of Domestic Livestock, L. Alderson, ed. Wallingford, U.K.: CAB International.

Ruane, J. 1992. The Global Data Bank for Domestic Livestock. Pp 49–54 in African Animal Genetic Resources: Their Characterization, Conservation, and Utilization. Addis Ababa, Ethiopia: International Livestock Center for Africa.

Schultz, R. D., L. S. Adams, G. Letchworth, D. E. Sheffy, T. Manning, and B. Bean. 1982. A method to test large numbers of bovine semen samples for viral contamination and results of a study using this method. Theriogenology 17:115–123.

Selander, R. K., and D. W. Kaufman. 1973. Genetic variability and strategies of adaptation in animals. Proc. Natl. Acad. Sci. USA 70:1875–1877.

Setshwaelo, L. L. 1990. Live animal conservation projects in Africa. Pp. 135–141 in Animal Genetic Resources: A Global Programme for Sustainable Development, G. Weiner, ed. FAO Animal Production and Health Paper No. 80. Rome, Italy: Food and Agriculture Organization of the United Nations.

Sharp, P. M. 1987. Molecular genetic distances among animal populations—Estimation and application. Paper presented at the 38th Annual Meeting of the European Association for Animal Production, September 27–October 1, 1987, Lisbon, Portugal.

da Silva, M. 1990. Programmes for live animal preservation for Latin America. Pp. 118–126 in Animal Genetic Resources: A Global Programme for Sustainable Development, G. Wiener, ed. FAO Animal Production and Health Paper No. 80. Rome, Italy: Food and Agriculture Organization of the United Nations.

Simon, D. 1990. The global animal genetic data bank. Pp. 153–166 in Animal Genetic Resources: A Global Programme for Sustainable Development, G. Wiener, ed. FAO Animal Production and Health Paper No. 80. Rome, Italy: Food and Agriculture Organization of the United Nations.

Singh, E. L. 1987. The disease control potential of embryos. Theriogenology 27:9–20.

Smith, C. 1984a. Estimated costs of genetic conservation of farm animals. Pp. 21–30 in Animal Genetic Resources Conservation and Management, Data Banks and Training. FAO Animal Production and Health Paper No. 44/1. Rome, Italy: Food and Agriculture Organization of the United Nations.

Smith, C. 1984b. Genetic aspects of conservation in farm livestock. Livestock Prod. Sci. 11:37–48.

Smith, C., T. H. E. Meuwissen, and J. P. Gibson. 1987. On the use of trans-genes in livestock improvement. Anim. Breeding Abstr. 55(1):1–10.

Society for the Advancement of Breeding Researches in Asia and Oceania (SABRAO). 1980. Proceedings of SABRAO Workshop on Animal Genetic Resources in Asia and Oceania. Tsukuba, Japan: Tropical Agriculture Research Center.

Soller, M., and J. S. Beckmann. 1983. Genetic polymorphism in varietal identification and genetic improvement. Theor. Appl. Genet. 67:25–33.

Somes, R. G., Jr. 1987. Linked loci of the chicken *Gallus gallus*. Genet. Maps 4:422–429.

Southern, E. M. 1975. Detection of specific sequences among DNA fragments separated by gel electrophoresis. J. Mol. Biol. 98:503–517.

Springmann, K., W. Schutz, and H. Krausslich. 1987. Factors affecting the utilization of genetic resources stored in an embryo bank. Paper presented at the 38th Annual Meeting of the European Association for Animal Production, September 27–October 1, 1987, Lisbon, Portugal.

Standing Committee on Cattle Blood Groups. 1985. Notice from a standing committee on cattle blood groups. J. Anim. Blood Groups Biochem. Genet. 16:249–252.

Stormont, C. 1959. On the application of blood groups in animal breeding. Pp. 206 in Proceedings of the Tenth International Congress on Genetics. Toronto, Canada: University of Toronto Press.

Tanksley, S. D., N. D. Young, A. H. Paterson, and M. W. Bonierbale. 1989. RFLP mapping in plant breeding: New tools for an old science. Bio/Technology 7:258–264.

Theilmann, J. L., L. C. Skow, J. F. Baker, and J. E. Womack. 1989. Restriction fragment length polymorphisms for growth hormone, prolactin, osteonectin, α crystallin, γ crystallin, fibronectin and 21–steroid hydroxylase in cattle. Anim. Genet. 20:257–266.

Thomas, D. L. 1991. Improvement of Prolificacy of U.S. and Israeli Sheep Populations Through Inclusion of the F Gene of the Booroola Merino. Bet Dagan, Israel: United States-Israel Binational Agricultural Research and Development Fund.

Trail, J. C. M. 1984. Animal genetic resources data banks. Pp. 124–127 in Animal Genetic Resources Conservation by Management, Data Banks and Training. FAO Animal Production and Health Paper No. 44/1. Rome, Italy: Food and Agriculture Organization of the United Nations.

Trail, J. C. M., G. D. M. D'Ieteren, and A. J. Teale. 1989. Trypanotolerance and the value of conserving livestock genetic resources. Genome 31:805–812.

Turton, J. D. 1984. The Commonwealth Bureau of Animal Breeding and Genetics and the provision of information for data banks on animal genetic resources. Pp. 142–147 in Animal Genetic Resources Conservation by Management, Data Banks and Training. FAO Animal Production and Health Paper No. 44/1. Rome, Italy: Food and Agriculture Organization of the United Nations.

U.S. Department of Agriculture (USDA). 1976. Handbook of Agricultural Charts. USDA Handbook No. 504. Washington, D.C.: U.S. Government Printing Office.

USDA. 1990. Handbook of Agricultural Charts. USDA Handbook No. 689. Washington, D.C.: U.S. Government Printing Office.

USDA. 1991. Animal welfare legislation alters European egg production. World Agric. 63:16–17.

Van der Zwalmen, P., K. Touati, F. J. Ectors, A. Massip, J. F. Beckers, and F. Ectors. 1989. Vitrification of bovine blastocysts. Theriogenology 31:270.

Vassart, G., M. Georges, R. Monsieur, H. Brocas, A. S. LeQuarré, and D. Christophe. 1987. A sequence in M13 phage detects hypervariable minisatellites in human and animal DNA. Science 235:683–684.

Wagner, T. E. 1986. Introduction and regulation of cloned genes for agricultural livestock improvement. Pp. 151–161 in Genetic Engineering of Animals: An Agricultural Perspective, J. W. Evans and A. Hollaender, eds. New York: Plenum.

Wainscoat, J. S., A. V. S. Hill, A. L. Boyce, J. Flint, M. Hernandez, S. L. Thein, J. M. Old, J. R. Lynch, A. G. Falusi, D. J. Weatherall, and J. B. Clegg. 1986. Evolutionary relationships of human populations from an analysis of nuclear DNA polymorphisms. Nature 319:491–493.

Watkins, P. C. 1988. Restriction fragment length polymorphisms (RFLP): Applications in human chromosome mapping and genetic disease research. Biotechniques 6(4):310–322.

Wezyk, S. 1990. Programmes for preservation of breeds in Eastern Europe. Pp. 127–134 in Animal Genetic Resources: A Global Programme for Sustainable Development, G. Wiener, ed. FAO Animal Production and Health Paper No. 80. Rome, Italy: Food and Agriculture Organization of the United Nations.

Whittingham, D. G., S. P. Leibo, and P. Mazur. 1972. Survival of mouse embryos frozen to –196°C and –269°C. Science 178:411–414.

Wiener, G., ed. 1990. Animal Genetic Resources: A Global Programme for Sustainable Development. FAO Production and Health Paper No. 80. Rome, Italy: Food and Agriculture Organization of the United Nations.

Willham, R. L. 1982. Historic development of the use of animal products in human nutrition. Pp. 3–17 in Animal Products in Human Nutrition, D. C. Beitz and R. G. Hansen, eds. New York: Academic Press.

Wilson, E. O., ed. 1988. Biodiversity. Washington, D.C.: National Academy Press.

Womack, J. E. 1987. Comparative gene mapping: A valuable new tool for mammalian developmental studies. Devel. Genet. 8:281–293.

Womack, J. E., and Y. D. Moll. 1986. Gene map of the cow: Conservation of linkage with mouse and man. J. Hered. 77:2–7.

Zabeau, M., and R. J. Roberts. 1979. The role of restriction endonucleases in molecular genetics. Pp. 1–63 in Molecular Genetics III, J. H. Taylor, ed. New York: Academic Press.

Zeuner, F. E. 1963. A History of Domesticated Animals. New York: Harper and Row.

APPENDIX

A

Global Status of Livestock and Poultry Species

Ian L. Mason and Roy D. Crawford

Unlike endangered wild species, none of the 40 or so animal species that have been domesticated for agricultural use is in serious jeopardy of extinction. Major species, such as cattle and pigs, are actually increasing in number as the food needs of a growing human population expand (Table A-1). Even those species of little importance on a global scale, such as reindeer and camel, are reasonably secure as long as people continue to live and maintain animals in the often inhospitable environments to which those species have adapted.

Breeds and native populations are the reservoirs of genetic diversity within species. Although livestock and poultry species are not at risk of extinction, a substantial—and, for some species, growing—number of breeds within those species are declining in population and size. Losses generally occur because the breeds no longer compete effectively within their respective production-market environmental niche. The extent to which the losses will actually reduce genetic variation within the species is not known nor easily measured.

Among the developed regions, considerable knowledge has been gained about the status and trends of livestock breeds in Europe,

---

Ian L. Mason is an animal breeding consultant and was, until recently, an animal production officer with the Animal Production and Health Division, Food and Agriculture Organization of the United Nations, Rome, Italy. Roy D. Crawford is professor emeritus of animal and poultry genetics at the University of Saskatchewan, Canada.

TABLE A-1 Population Sizes of Major Livestock and Poultry Species, by Region, 1988

Region or Country	Species						
	Cattle (000)	Buffalo (000)	Sheep (000)	Goats (000)	Horses (000)	Pigs (000)	Chickens (000)
Developed (predominantly)							
North America[a]	111,054	—	11,493	1,676	11,061	53,693	1,647,000
(United States)	(98,994)	—	(10,774)	(1,650)	(10,720)	(42,845)	(1,540,000)
Europe	124,780	367	142,081	12,470	4,340	190,412	1,280,000
Soviet Union	120,593	320	140,783	6,400	5,885	77,403	1,129,000
Oceania	32,122	—	228,983	2,050	579	5,365	73,000
Developing (predominantly)							
Central and South America	309,787	1,109	118,323	34,940	22,653	78,510	1,240,000
(Brazil)	(134,133)	(1,100)	(20,000)	(11,000)	(5,850)	32,700	(550,000)
Africa	181,190	2,600	199,599	166,993	3,655	13,057	828,000
Asia[b]	384,057	132,531	331,566	295,847	17,119	404,963	4,019,000
(India)	(193,000)	(72,000)	(51,684)	(105,000)	(953)	(10,300)	(260,000)
(China)	(73,963)	(20,858)	(102,655)	(77,894)	(10,691)	(334,862)	(1,849,000)
World total	1,263,584	136,927	1,172,828	520,376	65,292	823,403	10,216,000

[a]United States and Canada.
[b]Includes the Middle Eastern countries.

where a large number of local breeds were established and documented in the nineteenth and twentieth centuries. Both livestock producers and animal scientists have been concerned about the loss of breeds and have made concerted efforts to measure trends (Maijala et al., 1984) and to conserve minor breeds (Bowman, 1974).

North America and Oceania have no indigenous mammalian livestock breeds. Thus, if the numbers of imported breeds are small and declining, there is little cause for concern as long as the parent population from which they were drawn remains viable. However, populations can be identified as unique to these regions, and they merit consideration for conservation efforts.

Developing regions tend to have neither well-characterized breeds nor census data from which to derive population trends. Generally, little is known about the effective population sizes for most breeds, much less the degree of genetic diversity within and among breeds. It is in the developing regions that many of the more unusual and, perhaps, useful breeds are found. Unfortunately, they are the breeds that may be in greatest danger of genetic erosion.

This appendix provides an overview of the major livestock and poultry species in terms of total population sizes, global distributions, and transitional patterns in the structure of subpopulations. It is not intended to provide a thorough or exhaustive list of breeds, nor can it fully describe the extent of ongoing changes in distribution or losses. Rather, it is intended to provide a sense of (1) the overall diversity in terms of breeds or types of stocks in different regions of the world and (2) the impact on this diversity of such factors as the international movement of germplasm, increased popularity or efficiency of certain breeds, and the synthesis of new breeds. The discussion is largely based on reports by Ian Mason (1984, 1988) about cattle, goats, pigs, horses, sheep, and buffalo and by Roy Crawford (1984, 1990) about poultry.

CATTLE

Two major types of cattle—humpless and humped—comprise the world cattle population. They were originally named as separate species, *Bos taurus* and *Bos indicus*, but they are interfertile and now generally regarded as a single species. The former evolved and today predominates in temperate regions around the Mediterranean, in Europe, and across northern Asia to Japan. They were taken to the Americas, Australia, New Zealand, and South Africa by European colonists. *Bos indicus*, noted for their thoracic hump, are commonly called zebu. They predominate in southern Asia, from Saudi Arabia through southern and central China and Indonesia, and in tropical Africa. Intermating among humped and humpless cattle along the extended borders of their range has produced cervicothoracic humped types, such as the Sanga breeds of eastern and southern Africa.

Europe

The Friesian (Black Pied Lowland) is the dominant breed in western Europe. (Friesian is used here as a shorthand term for the Black Pied dairy breeds, including the Holstein.) Major populations of Simmental, Swiss Brown, and Red Pied Lowland also exist. Nevertheless, many of the original native breeds are still maintained, although many are rare or nearly extinct (Table A-2). In most of western Europe, private or governmental bodies operate conservation programs for rare breeds.

In Scandinavia, imported Friesians and Jerseys and derivatives of the Shorthorn and Ayrshire have all but eliminated the local breeds. Only the Danish Red and the Icelandic remain as major breeds; the Faroes and Finnish as minor breeds; and the Swedish Mountain, Blacksided Trondheim, and Telemark (both the latter in Norway) as rare breeds. A conservation program is monitoring these breeds.

In central Europe, the Swiss Brown and Simmental from Switzerland, the Red Pied Lowland (in Germany and Poland), and the Friesian have supplanted most of the local breeds. However, there still remain major populations of Pinzgauer and Tyrol Grey in Austria

TABLE A-2 Number and Status of Cattle Breeds in Western Europe

Region or Country	Status				
	Major	Minor	Rare	Nearly Extinct	Recently Extinct
Belgium	3	—	—	—	—
British Isles	10	4	8	3	2
France	7	3	6	5	2
Italy	8	6	9	3	2
Malta	—	—	—	1	—
Netherlands	—	1	2	—	—
Portugal	3	5	3	—	—
Spain	6	8	2	4	3
Total (excluding overlaps)	36	26	30	14	9

NOTE: Excludes Friesian, Brown, Simmental, Red Pied Lowland, and their derivatives.

and of Angler and Gelbvieh in Germany, and minor populations remain of Polish Red in Poland and of Hérens in Switzerland. In Czechoslovakia, all the native red (or color-sided) breeds have disappeared since World War II.

In southeastern Europe, the Simmental and Swiss Brown, to a small extent the Pinzgauer, and, more recently, the Danish Red and the Friesian have almost completely replaced the local breeds. The Simmental has given rise to the several national pied or red pied breeds. The Grey Steppe is nearly extinct in Romania and Yugoslavia, and it is rare in Albania, Bulgaria, Greece, and Hungary. The native *brachyceros* (shorthorn type of *Bos taurus*) breeds are nearly extinct in Romania and rare in Bulgaria and Greece; only in Albania and Yugoslavia are there sizable populations of breeds whose origins are traceable to the shorthorn type.

The former Soviet Union (now the Commonwealth of Independent States) has more recognized cattle breeds than any other country in the world (Dmitriev and Ernst, 1989). With the specific exception of the Kalmyk in central Asia, however, the 30 major breeds are derivatives of black pied, brown, red, red pied (Simmental), or white-headed breeds of western and central Europe. Native breeds of minor importance include Azerbaijan Zebu, Central Asian Zebu, Georgian Mountain, and Mingrelian Red; rare and declining breeds include Dagestan Mountain, Estonian Native, Goryn of Byelorussia, Kazakh, Siberian White, Ukrainian Grey, and Yakut. Conservation herds have been formed for three breeds of the latter group. Several other breeds have recently become extinct.

Asia

In most of the Arab countries of southwest Asia, cattle are not of major importance. In Turkey, however, there are large populations of three local humpless breeds and smaller numbers of the imported Grey Steppe and Brown breeds. Syria also has its local humpless breed, as well as the declining Damascus zebu × humpless breed, which is similar to breeds in Cyprus, Egypt, and Lebanon. In Iraq four local breeds are of zebu or zebu × humpless origin. In Iran interest is reviving in the local breeds, and conservation programs are being developed for five local breeds that vary from zebu in the east to humpless in the west.

As for the Indian subcontinent, there are 23 major and 6 minor recognized zebu breeds in India, 6 major and 2 minor in Pakistan, and 1 each in Bhutan and Sri Lanka. Only about 20 percent of cattle in the subcontinent can be classified into recognized breeds; the rest are desi—that is, local, village, or unimproved cattle. Two breeds in India are categorized as rare (Umblachery and Punganur of Tamil Nadu) and two more (Sahiwal and Ongole) should be the focus of conservation programs. Extensive crossing with European breeds has produced five named breeds of crossbred origin; the crossbreeding programs tend to choose the best Indian breeds as foundation stock, which has led to the decline in the number of pure Sahiwals. Two nineteenth-century derivatives of European crossbreeding, the Taylor of Bihar and the Hatton of Sri Lanka, are nearly extinct. The unique mithun (*Bos frontalis;* domesticated gaur) is found in Bhutan and neighboring countries to the east.

In Japan, the original breeds have been displaced by crossing with European breeds to form four improved Japanese breeds. Elsewhere in the Far East, however, the local cattle remain and crossbreeding has only just begun. China has humpless cattle in the north, pastoral areas of Iko (Mongolian) and in the west (Kazakh and Tibetan) and zebus in the south (but two out of the three breeds are now rare); in central China there are four major and four minor breeds with some zebu genes. Korea still has its native humpless cattle. Farther south, Burma, Indonesia, Kampuchea, Malaysia, the Philippines, Thailand, and Vietnam each has its own zebu landrace. In addition, Indonesia has the unique Bali cattle (*Bos javanicus*), which are descended from the banteng, and Madura cattle, which are intermediate between Bali cattle and Javanese zebu.

Africa

In North Africa, European breeds have had a great influence, but the native humpless breeds (Brown Atlas and Egyptian) remain dominant. The Oulmès Blond (Morocco) and the Cape Bon Blond (Tunisia) are two local derivatives. The Thibar is a newly developed breed in Tunisia. Some Libyan cattle remain; they are intermediate between the Egyptian breed and the Brown Atlas.

In West Africa, there has been very little crossing with imported breeds because they generally adapt poorly to local disease and climatic problems. There are five major local types: zebus, N'Dama, West African Shorthorn, zebu × humpless crossbreeds, and Kuri.

In the northern part of West Africa, there are 11 recognized breeds of zebu. These are the cattle that suffer from the periodic Sahelian droughts. In the southerly (coastal) areas, where the tsetse fly and associated trypanosomiasis disease are prevalent, are the two humpless trypanotolerant breeds: N'Dama in the west and West African Shorthorn in the east and south. Between zebu and humpless at least four stabilized breeds of crossbred origin have developed. This crossbreeding is putting a strain on the West African Shorthorn, however; the zebu or zebu-cross is preferred because of its larger size, despite its reduced (or lack of) tolerance to trypanosomiasis. Of the several varieties of West African Shorthorn, those in Gambia, Guinea-Bissau, and Cameroon are either extinct or nearly so, and those in Togo, Benin, and Liberia are declining. Because its productivity per unit of body weight is higher than that of other breeds in tsetse areas, the West African Shorthorn needs conservation. There are still sizable populations in the Ivory Coast (Baoulé) and Nigeria (Muturu). The fifth West African breed, the humpless Kuri of Lake Chad, is unique and its status should be monitored.

In East Africa, the original humpless cattle have disappeared (unless they are represented by the rare Sheko breed of southwest Ethiopia); they were displaced first by the Sanga and then by zebu from India. There are now recognized Sanga breeds in Ethiopia, Sudan, Uganda, Zaire, and Tanzania; in addition, there are at least 20 zebu breeds in those countries and in Kenya, Mozambique, and Madagascar, as well as 5 zebu-Sanga intermediates. The only endangered breeds appear to be the above-mentioned Sheko, two breeds of European × zebu origin (the Mpwapwa in Tanzania and the Renitelo in Madagascar), and two "breeds" in Madagascar (the Baria and the Rana) about which recent information is lacking.

In southern Africa, there are many Sanga breeds in Angola, Botswana, Lesotho, Namibia, South Africa, Swaziland, Zaire, Zambia, and Zimbabwe. No breeds are recorded as rare, but there are some interesting new breeds in comparatively small numbers in Botswana, Namibia, and South Africa, all based on a European × Sanga foundation.

The Americas

America has no truly indigenous cattle; cattle of Spanish or Portuguese origin (Criollo or Crioulo) have been acclimatizing for nearly 500 years, however, and they have some claim to being called native. In the temperate areas (Argentina, Chile, Uruguay, southern Brazil, northern Mexico, and the United States) the native cattle have been replaced by more productive European breeds, and in the tropical areas they are giving way to the zebu. The improvement in zebu × Criollo crosses is wrongly attributed to additive rather than to heterosis effects, so grading up continues and the pure Criollo is now rare. The largest populations are the Yacumeño of Bolivia, the Casanareño/Llanero of Colombia and Venezuela, and the mountain Criollos of the Andes. There are several other local breeds in Colombia but not in large enough numbers to be exploited commercially. Small Criollo populations are being used for meat production in Argentina, Bolivia, Colombia, Cuba, and Venezuela. Criollo cattle for milk production have been developed in Brazil, the Central American countries, the Dominican Republic, Colombia, Bolivia, and Venezuela. De Alba (1987) describes 31 Criollo ecotypes.

In the United States, the two breeds of Spanish origin are the Texas Longhorn, which has come back from near extinction to be a flourishing breed, and the nearly extinct Florida Cracker, which is now the object of private conservation efforts. In the United States and Canada, some interesting rare breeds are descended from early imports from other countries; these include the Dutch Belted, Milking Devon, and Canadienne.

In both North and South America, the most striking feature is the multitude of new breeds, either well established or in formation. In tropical America, the new breeds are mostly based on a zebu × European cross, but some are Criollo × European and others zebu × zebu. They include nine breeds in Brazil (beef, dairy, and dual purpose), two in Colombia, two in Cuba, and four in Jamaica. There is also the Senepol of the Virgin Islands, which is unique in that it has N'Dama as well as Red Poll genes. In the United States, all the new breeds are beef type; seven are based on a zebu cross and five have only European ancestors. Three breeds are said to have bison blood. In Canada, no zebu genes is present in the one new dairy and four new beef "breeds."

Australia

Like America, Australia is chiefly noteworthy for its new breeds. The Illawarra and Murray Grey breeds, of pure European origin, are of long-standing and major importance. In the past 60 years, at least eight new beef and two new dairy breeds have been formed from initial zebu × European crosses.

GOATS

Europe

In all of the countries of Europe except Norway and the southern peninsulas, the goat is decreasing in importance. All major breeds are derivatives of Swiss dairy goats, especially the Saanen, Toggenburg, and Chamois Colored, in that order. A few minor local breeds remain, including the Anglo-Nubian in England; the Corsican in France; the Pinzgau in Austria; the Carpathian in Poland and Romania; and the Appenzell, Grisons Striped, Valais Blackneck, and Verzasca in Switzerland. There are also several rare breeds, mostly kept by fanciers or in conservation herds. These include the Bagot, Old English, Golden Guernsey, and Irish in the British Isles; the Dutch Dwarf, Dutch Landrace, and Dutch Pied in the Netherlands; the Catalan, Massif Central, Poitou, Provençal, and Rove in France; the St. Gallen in Switzerland; and the Peacock goat in Germany.

The importation of goats raised for fiber is a recent and growing interest in many European countries. The use of the imported animals to upgrade local goats for fiber production has further jeopardized the survival of some populations.

In Scandinavia, the Norwegian goat has absorbed the declining native goats of Sweden and, to a lesser extent, Finland to form the flourishing Nordic breed. The Icelandic breed, which was down to only 200 animals, has been saved through a state program initiated in the 1970s.

The number and status of goat breeds in southern Europe are shown in Table A-3. The table excludes the Swiss derivatives in Italy, which are among the major breeds in that country. The large number of minor and rare breeds reported for Italy reflects recent identification of several local breeds in what was hitherto considered a heterogeneous population and, to a great extent, still is.

In the Commonwealth of Independent States, dairy goats are mostly restricted to individual holdings; the only recognized breeds are the Saanen derivatives in northern European Russia (Russian White and Gorki) and the native Mingrelian in Georgia. The state and collective farms in the Don basin and in central Asia keep large herds of down (cashmere) and mohair goats of several breeds, most of them of recent formation from crosses onto the local coarse-haired goats. The latter are found in small numbers in central Asia and in Transcaucasia.

Asia

In southwest Asia, there are many interesting and important goat breeds. These include the Angora and Anatolian Black of Turkey, the Mamber and Damascus of Syria, the Israeli Saanen (the only European influence), and the Iraqi. These are the breeds listed as "major" in Table A-3. The Damascus goat of Syria, a dairy breed, is in need of conservation and improvement. The minor breeds of southwest Asia may also be numerically and economically important, but they have not been adequately described, except for those in Turkey.

In the Indian subcontinent, there are very many goats, and a large proportion have been classified into recognized or local breeds (Table

TABLE A-3 Number and Status of Local Goat Breeds

Region or Country	Status			
	Major	Minor	Rare	Nearly Extinct
Southern Europe				
Albania	1	—	—	—
Bulgaria	—	3	—	—
Greece	1	—	—	—
Italy	4	11	7	4
Portugal	3	2	—	—
Spain	5	5	—	—
Yugoslavia	1	1	—	—
Southwest Asia				
Afghanistan	—	3	—	—
Cyprus	—	2	—	—
Iran	—	6	—	—
Iraq	2	2	—	—
Israel	1	2	—	—
Syria	2	—	—	—
Turkey	2	2		
Saudi Arabia	—	1	—	—
Yemen (North)	—	5	—	—
Indian subcontinent				
Bangladesh	1	—	—	—
India	18	2	—	—
Kashmir	1	6	—	—
Nepal	1	3	—	—
Pakistan	13	5	—	—
Sri Lanka	—	2	—	—

A-3). These include dairy, meat, hair, and cashmere producers. The only endangered breeds in India are the Jamnapari, whose purebred numbers have fallen to 5,000, and the Barbari, of which there are 30,000 purebreeds.

China rivals India in the number and variety of its native goats. Many breeds have been briefly described, and they can be tentatively classified into 15 major and 14 minor breeds. They include meat and cashmere breeds. In addition, eight new dairy breeds are being formed by crossing native goats with European breeds, chiefly Saanen.

In the Far East, there are small, local goats in Korea, Indochina, the Philippines, Indonesia, Malaysia, and Thailand. The best known is the Katjang of Malaysia and Indonesia. Japan's native breeds have been reduced to two rare populations, one in a protected herd and the other feral. In southeast Asia, there is considerable crossing with the Jamnapari of India.

Africa

In North Africa, the goats in the east (Egypt) are clearly related to the long-haired, long-eared black breeds of the Middle East (Anatolian Black, Mamber, Iraqi). Farther west they become more heterogeneous, and "breeds" have not been described (except for the Mzabite of Algeria). The Zaraibi dairy goat of Egypt appears to be declining in number, and its status should be monitored.

The goats of West Africa fall into two main groups—the West African Dwarf, found toward the coast, and the Sahelian, found farther inland. The West African Dwarf is remarkably uniform from Senegal to Angola, but the Sahelian is divided into many local "breeds." The Maradi of Niger (a breed equivalent to the Red Sokoto of Nigeria) is a useful producer of "Morocco" leather.

In East Africa, two breeds have been described in Ethiopia (with many local varieties), four in the Sudan, one major and three minor in East Africa proper, three in Somalia, and one in Mauritius. In southern Africa, there is one improved meat breed (the Boer of South Africa) and flourishing native populations in all other countries of the area.

The Americas

The original Spanish, Portuguese, and West African imports have given rise to the Criollo goats of Spanish America (with many national populations) and the Crioulo of Brazil. From the latter has developed the SRD (*sem raça definida,* without defined breed) with its four color varieties, which are being made into breeds. Interesting new imports into Brazil are the Bhuj from India, the Mamber from Syria, and the Anglo-Nubian and other European breeds.

In the United States, several new breeds of diverse origin have been formed, including Lamancha (Mexico), American Pygmy and Nigerian Dwarf (West Africa), Oberhasli (Switzerland), Spanish Angora, and Tennessee Fainting (India). Perhaps the most interesting goat genetic resource in the Americas is the series of feral populations found on at least five Pacific islands. All are in danger from extermination policies designed to save the native plant and animal communities of the islands.

PIGS

Europe

In western and central Europe, the spread of the highly productive Large White (also called Yorkshire) and Landrace breeds has all but eliminated the many local breeds that were designed for extensive management or were lard producers and so had lower meat productivity in intensive management systems (see Table A-4).

The one major native breed is the Piétrain of Belgium. There are some new "breeds" in Britain, Germany, the Netherlands, and elsewhere, but they are commercial hybrids of existing breeds. Scandinavia is the home of the Landrace breeds. The only other native breeds are the rare Danish Black Pied and the extinct Old Swedish Spotted.

In southeast Europe, there still appear to be some native breeds (Table A-4). Many of these have the influence of west European breeds—especially the Berkshire and Large Black.

In the Commonwealth of Independent States, there are 19 recognized breeds, but 11 are Large White or its derivatives (some with

TABLE A-4 Number and Status of Local European Pig Breeds

	Status				
Region or Country	Major	Minor	Rare	Nearly Extinct	Recently Extinct
Western and central Europe					
Benelux[a]	1	—	—	—	—
British Isles	—	—	1	7	6
France	—	—	1	4	16
Germany	—	—	1	3	6
Hungary	—	—	—	1	—
Italy	—	1	2	4	12
Poland	—	2	1	1	2
Portugal	—	4	—	1	—
Spain	—	1	—	2	8
Southeast Europe					
Albania	—	1	—	—	—
Bulgaria	—	2	—	1	—
Romania	—	3	—	—	—
Yugoslavia	—	4	2	—	4

NOTE: Excludes Large White, landrace, and their derivatives.

[a]Belgium, Luxembourg, and the Netherlands.

Landrace genes) and the others also have Berkshire or Middle White genes. The only truly indigenous breed is the rare Kakhetian of Georgia. Even the 20 recently extinct breeds (or "breed groups") are almost entirely (at least 17) breeds of mixed origin. New breeds are chiefly commercial hybrids of the Landrace, Large White, and other breeds.

Asia

There are no pig breeds in the Muslim countries of southwest Asia, but there are three native breeds in India and one in Sri Lanka. China has more pig breeds than any other country in the world. The Chinese divide them into about 50 breeds, with over 100 varieties, but they are also classified into 15 major breeds (Cheng Peilieu, 1984) and 47 minor breeds. Many of the breeds are of interest because of their high prolificacy; all are adapted to low-quality diets, and many are primarily lard producers. There are also 16 new breeds formed from Large White, Berkshire, Middle White, and Landrace crossed with local breeds. The local breeds are not in danger, however, owing to sheer numbers. In Taiwan, the situation is quite different; the local breeds have been almost entirely replaced by improved breeds from the United States and Europe to suit the intensive husbandry systems that have been developed.

In southeast Asia, native breeds are reported as follows: Indonesia, four; Malaysia, three; Philippines, one; Thailand, three; Vietnam, eight. In Malaysia and Thailand, and probably elsewhere, some are of Chinese origin. Little is known of their status or performance. There are also breeds of crossbred origin (from North American and British breeds) in the Philippines and Vietnam. There are native breeds in Korea and in Papua New Guinea.

Africa

There are few, if any, pigs in Muslim North Africa and only two native breeds have been recorded elsewhere—the West African and the Bantu of South Africa. The former is probably of Portuguese origin and the latter of Chinese. Both may contain unique adaptive genes for disease and other environmental stresses.

The Americas

There are many locally adapted pigs in Latin America; they are generally called Criollo. Four major and six minor breeds are reported from Brazil, and there are 10 local varieties in the tropical countries between Mexico and Guyana. Some of the local varieties are becoming rare. Native breeds have little place in the economy, except in Brazil; in Argentina, Chile, Mexico, and Uruguay, they have no place at all. The imported breeds chiefly used are Poland China, Duroc, Hampshire, and Large White.

The United States is unique among Western countries in having its own breeds of pigs. In addition to Landrace and Large White, there are five major and two minor breeds, two more are nearly extinct, and four are recently extinct (not counting the short-lived synthetic breeds). There are also feral populations that may be of interest and several breeds of miniature pigs. Canada has its own new breeds, Lacombe and Managra, which are of multibreed origin.

Australia and New Zealand

Both countries have large feral populations. New Zealand has the Kunekune (Maori) pig, thought to have originated from pigs brought in the early sailing ships.

HORSE

Europe

This overview does not include Arabs, Thoroughbreds, or Trotters because they can be considered "international" improved breeds rather than indigenous types. There has been so much crossing with these breeds among light horses, however, that the distinction between breeds is sometimes quite arbitrary. Similarly, nearly all draft breeds carry genes of British, French, or Belgian breeds.

In western Europe, draft breeds continue to be of importance only in France, where they are converted into meat breeds. In general, the only flourishing breeds are trotters, race horses, and riding breeds. The only truly indigenous breeds are to be found among the ponies (see Table A-5).

In Scandinavia, there are still major (Fjord and Icelandic Pony) and minor (Faeroes Pony and Finnish) indigenous breeds, but most breeds are derivatives of west European breeds. Two other local breeds, Gotland Pony and Northland (Norway), are rare.

Austria still has two important local breeds, the Haflinger and the Noric, which extend into Italy and Germany, respectively. The famous Kladruby horse of Czechoslovakia is nearly extinct, and the local Gurgul of the Slovakian Carpathians is probably also. In Germany, the draft breeds (of Belgian and Suffolk origin) are now rare,

TABLE A-5 Number and Status of Horse Breeds in Selected West European Countries, by Type

		Status			
Region or Country	Type of Horse	Major	Minor	Rare	Nearly Extinct
Benelux[a]	Draft	—	2	—	—
	Riding	—	1	1	1
British Isles	Draft	—	—	4	—
	Riding	1	2	0	—
	Pony	3	3	4	1
France	Draft	4	2	1	—
	Riding	—	1	—	—
	Pony	—	1	3	—
Italy	Draft	—	1	—	—
	Riding	—	2	5	1
	Pony	—	1	2	—
Portugal	Riding	—	1	—	—
	Pony	—	3	—	—
Spain	Draft	—	1	—	—
	Riding	1	—	—	—
	Pony	—	3	—	1

NOTE: Excludes Arab, Anglo-Arab, Thoroughbred, and Trotter breeds.

[a]Belgium, Luxembourg, and the Netherlands.

and the local warmblood breeds have been combined into the German Riding Horse (West Germany, formerly) and the Edles Warmblut (East Germany, formerly). Two local ponies are almost extinct. In Hungary, the Hungarian Draft (of west European origin) and the Hungarian Halfbred (of Thoroughbred origin) have eliminated the native Hungarian horse. In Poland, the light horses (mostly of English Halfbred origin) have been combined into two regional breeds. There are also several draft breeds of west European origin. The unique breeds are the rare Hutsul and Polish Konik. Switzerland has the light draft Freiberg breed, which is of French origin.

In southeast Europe, the situation is fluid. The Balkan ponies are still important in Albania and Yugoslavia, less so in Greece, and almost extinct in Bulgaria. Other major breeds are the Nonius, the three Halfbred breeds in Bulgaria, and the Lipitsa of Yugoslavia. Bulgaria and Yugoslavia have heavy draft breeds, and Romania has light draft breeds, which are based on imports.

The Commonwealth of Independent States has 18 major, 9 minor, 16 rare, and 17 recently extinct breeds. Except for the local Altai and Yakut ponies, all the major light breeds are based on the Arab or Thoroughbred or their derivatives; the major draft breeds are based on Belgian, British, and French breeds. The minor breeds include four native breeds and five native breeds improved by crossbreeding. All the rare breeds, except one, are indigenous breeds. They include six north Russian pony breeds, two Siberian ponies, and four ponies from the Caucasus. The recently extinct include five native breeds.

Asia

Horses in southwest Asia are chiefly Arabs or of Arab origin. In addition, there are three named breeds of light horses in Iran, three in Turkey, and three in Afghanistan. The native horse of Turkey, the Anatolian Pony, is common. In Iran, the Caspian Pony is a recently revived rare breed.

In the Indian subcontinent, horse breeds are almost confined to the north. The Deccani breed is nearly extinct. There are three major and two minor breeds in India (two of them ponies) and two major, two rare, and one nearly extinct breed in Pakistan. In Nepal, Bhutan, and northeast India, the Bhotia Pony is similar to the Tibetan Pony.

In China, the few major breeds are in the north and west. The dominant breeds are the Mongolian and Tibetan ponies, but there are four other ponies in the southwest and three light horse breeds with Russian genes in the north.

In southeast Asia, every country has a named population of ponies (Indonesia has no less than seven) that are collectively called the "South-East Asia Pony." The native Korean and Japanese horses are included in the same group. The status of most of these populations is unknown, but all of the isolated populations in Japan are rare or nearly extinct.

Africa

The horse breeds of North Africa are the Barb in the Maghreb and the Egyptian, an Arab derivative. In West Africa, the common light horses are the West African Barb in the west and the West African Dongola in the east. Both have varieties and derivatives. There are also seven named breeds of ponies; the four in the west may be degenerate Barbs, but the three in the east (Bhirum, Koto Koli, and Kirdi) are genuine ponies. It would be interesting to know whether they are resistant to trypanosomiasis.

The horse breeds of East Africa are the Dongola and Western Pony of Sudan and the Abyssinian horse. The Somali Pony is rare. In southern Africa, there are two or three local breeds of horse or pony that are struggling to survive, and the Basuto Pony is nearly extinct.

The Americas

Latin America has its Criollo (or Crioulo) breed or breeds. Some are improved Criollos (such as the Argentine), some are mountain ponies, and some are improved breeds with foreign genes (such as the Campolino and Mangalarga of Brazil). There is also the miniature Falabella Pony of Argentina, which is of a different origin.

The United States has 8 major, 12 minor, and 8 rare breeds. All are light horses or ponies; some breeds are based on performance, but many are primarily color types and some are feral or of feral origin. Canada has one native rare breed (Canadienne), one local pony, and one feral pony.

Australia

Riding horses are used in Australian agriculture as much as anywhere in the world, but breeds have not been developed. They are all the Australian Stock Horse.

SHEEP

Europe

In western Europe, there are still very many breeds of sheep, although the decline in numbers (except in Spain and Britain) and the spread of more productive breeds (for example, hill breeds in Britain, milk breeds in France, and Merino breeds in Italy) have led to a large tally of rare and recently extinct breeds (Table A-6). Active conservation societies are found in Britain, France, Switzerland, and the Netherlands.

The situation is remarkably similar in all countries, but the wave of extinction has not yet reached Spain and Portugal. The large number of rare or extinct breeds indicated for Italy in the table reflects a recent survey that identified new "breeds" among local populations. The breeds that remain in western Europe include all the European types: fine, medium, and coarse wools; prolific, meat, and milk sheep; mountain, hill, and lowland types; and fancy breeds with spots or

TABLE A-6 Number and Status of European Sheep Breeds

Region or Country	Status					
	Major	Minor	Rare	Nearly Extinct	Recently Extinct	New
Western Europe						
Benelux[a]	1	2	3	1	3	3
British Isles	16	23	8	4	3	11
France	20	14	7	4	11	1
Italy	16	9	13	10	8	1
Portugal	11	2	—	—	—	1
Spain	14	13	4	—	—	1
Central Europe						
Austria	1	—	1	2	—	—
Czechoslovakia	—	3	—	2	—	—
Germany	6	2	5	1	—	—
Hungary	1	—	1	1	—	—
Poland	1	11	2	—	1	—
Switzerland	1	3	—	4	—	—
Southeast Europe						
Albania	2	3	—	—	—	—
Bulgaria	2	3	15	—	3	11
Greece	5	14	9	—	1	—
Romania	2	1	—	—	—	5
Yugoslavia	7	11	5	—	—	—

[a]Belgium, Luxembourg, and the Netherlands.

four horns. Most of the new breeds incorporate genes from prolific or dairy breeds.

In Scandinavia, at least one native breed remains in each country; the Norwegian and Swedish Landraces can each be divided into three subbreeds. In addition, there are three Norwegian breeds of early British origin. The numbers are small but fairly stable.

In central Europe, the process of replacement of native breeds has gone much farther. This may not be clear from Table A-6, but with the exception of the East Friesian in Germany, all major breeds are of Merino or other west European origin, as are all but four of the minor breeds. The rare, nearly extinct, and completely extinct breeds, on the other hand, are with one exception true indigenous breeds. The Swiss conservation society has been very successful in finding and saving nearly extinct breeds.

In southeast Europe, the assault on the local breeds is well under way. The replacement of local breeds is proceeding most rapidly in Bulgaria and Romania, where large farms make rapid changes easy.

New breeds have been formed by crossing the native breeds with Merinos and west European meat or wool breeds. The process is continuing, but in Bulgaria the government has formed conservation flocks for some rare breeds. In Greece, the decline is due rather to a decline in the sheep industry, resulting in part from the greater attraction of tourism.

The large number of breeds in Bulgaria, Greece, and Yugoslavia, compared with Romania, is due to the fact that the Zackel (the common coarse-wooled Balkan sheep) is divided into 21 "breeds" in Bulgaria, 16 in Greece, and 19 in Yugoslavia, whereas in Romania only one Zackel breed is recognized. Thus, the great loss of breeds in the three countries probably does not represent a corresponding loss of genetic variation.

In the Commonwealth of Independent States, the situation is similar, but the breeds are much more diverse. There are 27 major, 25 minor, 7 rare, and 38 new breeds; 16 breeds have recently become extinct. Fourteen of the major breeds, seven of the minor, and all the new breeds originate from Merinos or from British mutton or wool breeds crossed onto local sheep. On the other hand, six of the rare breeds and all the recently extinct breeds are indigenous. Thus, the pressure on the local breeds is extreme, and because many are confined to farmers' individual holdings, it is difficult to get exact information.

Asia

Southwest Asia is a sheep area par excellence. There are large numbers of breeds but not much apparent variety because all have carpet wool (or very coarse hairy wool), and most are characterized by a fat tail. Afghanistan recognizes 8 breeds; Cyprus, 1; Iran, 16; Iraq, 3; Oman, 1; Saudi Arabia, 5; Syria, 2; Turkey, 12; and North Yemen, 9. The total number of breeds is smaller, however, because several breeds (notably Arabi, Awassi, and Baluchi) each extend into two or more countries. The numbers of all breeds appear adequate, and there are no records of rare or declining breeds. There has been very little crossing with foreign breeds, and the only notable new breed is the Turkish Merino.

The Indian subcontinent has a very large sheep population, varying from the fat-tailed, coarse-wooled breeds of Pakistan to the short-tailed hair sheep of southern India. There are 38 native breeds in India, 28 in Pakistan, 5 in Nepal, and 1 each in Sri Lanka and Bangladesh. In addition, there are about six new breeds in India based on Merino crosses, but they have not yet made any impact on the numbers of the native breeds, except in Kashmir, where the Kashmir Merino has reached large numbers. Endangered breeds are Gaddi, Bhakarwal, and Poonchi of Jammu and Kashmir (owing to crossbreeding); the Magra, Pugal, and Chokla breeds of Rajasthan (owing to shortage of feed and water in their arid habitat); and the Mandya of Karnataka (in part because of the spread of cultivation).

In China, there are six major and six minor breeds, chiefly in the pastoral areas of the west and north. They are all coarse-wooled breeds. Fourteen new breeds have been formed (or are being formed) by crossing with western breeds, chiefly Merinos. However, the new breeds do not yet seem to have made any impact on the numbers of the local breeds, especially the Mongolian, which is dominant in Mongolia and northeast China. The Hu and the Han are interesting prolific breeds in central China.

Southeast Asia is not sheep country, but there are small populations of the local breeds in Burma, Malaysia, and Thailand. Indonesia is the only country of the region with a large population of sheep, chiefly in Java. There are three breeds, all of interest because of their adaptation to a hot, humid environment, high reproductive rate, and mutton production.

Africa

Among the countries of North Africa, Morocco, Algeria, and southern Egypt have thin-tailed sheep; Tunisia, Libya, and northern Egypt have fat-tailed sheep. All are coarse-wooled breeds and by necessity adapted to a difficult environment. Crossing with improved breeds has had little effect on the general sheep population. Three native breeds can be counted in Algeria, six in Egypt, ten in Morocco, one in Tunisia, and one in Libya, but subdivision into breeds is flexible in Algeria and Morocco. Some interesting breeds in Morocco should be monitored; for example, the less coarse-wooled breeds of the Atlantic Coast and the prolific D'Man of the Sahara. A new breed is the Tunisian Milk sheep, of Sardinian origin.

West Africa has five major breeds with many varieties. Only one, the Macina of Mali, is wooled. The others are hair sheep—long-legged sheep in the Sahel and the so-called West African Dwarf in the Guinean (coastal) zone. The latter is trypanotolerant. Numbers appear to be stationary, and there has been almost no crossing with exotic breeds.

In East Africa, three breeds have been described in Ethiopia, but the country has not been covered, and several more breeds have not been adequately described. Sudan has two main breeds with many varieties and at least three minor breeds. Somalia has its famous Blackhead. Farther south there are well-defined populations of fat-tailed hair breeds. No attrition is reported, but in the highlands of Kenya, European breeds continue to displace the indigenous Masai sheep.

In southern and central Africa, each country has its population of local fat-tailed hair sheep, but in Zaire, Angola, and west central Africa, the numbers are few and those that are found are poorly described. In southern Africa, most of the local breeds have been replaced by European breeds, especially Merinos. The Blackhead Persian, of Somali origin, has taken the place of the local breeds as an adapted hair breed. The Nguni remains in Zululand, as well as in Swaziland and Mozambique, but the Africander is nearly extinct. Most new breeds in southern Africa are based on the Merino, but the Dorper is of Dorset-Persian breeding and a useful addition to improved meat breeds adapted to the dry tropics.

The Americas

The temperate areas of Latin America are entirely populated by European breeds, largely Merinos, but the tropical areas have large populations of "native" breeds, so-called because of centuries of adaptation to local conditions. The Criollo, of Spanish origin, is the major breed in the highlands from Mexico to Bolivia. There are many varieties but few have been described. In the tropical lowlands from Mexico to Brazil, hair sheep (originally from West Africa) have become increasingly important. There are at least four recognized breeds and many mixed and undescribed populations. They are also important on several Caribbean islands. There are prolific breeds on Barbados and the Virgin isles, but the numbers are small and the breeds need encouragement. There are other groups in the Bahamas, Cuba, and the Dominican Republic (among others).

In the United States, most of the major British breeds are represented, as well as many breeds from other countries. Many synthetic breeds were formed in the past. These can be classed as two major, five minor, six rare, two feral, and one nearly extinct. Canada also has two minor breeds. The process of breed formation continues; 11 new breeds in the United States, plus 2 in Canada, have been developed in the past 50 years from a very diverse selection of foundation breeds.

Australia and New Zealand

Besides the British breeds, the major breeds of local origin are the Australian Merino, Corriedale, New Zealand Romney, and Polwarth.

Through various combinations of these resources, 12 new breeds have been formed in Australia and 10 in New Zealand. In New Zealand, one aim has been to produce carpet-wool breeds by exploiting the coarse-wool mutations of the Romney.

BUFFALO

There are two main types of water buffalo: the Swamp and the River buffalo. The Swamp buffalo of southeast Asia extends from Assam to the Philippines and from southern China to northern Australia. They are primarily draft and secondarily meat animals. There is little apparent variation, except in size and horn length. Their color is slate grey, but there are some white variants in Thailand and some pied in Indonesia. No breeds have been defined.

The River buffalo extends from India westward to Egypt and Italy. These buffalo are black or dark grey, have a variety of horn forms, and are primarily dairy animals. They have been developed into a series of seven recognized breeds in India and Pakistan. At least five named varieties are among the prevailing *desi* population of central and southern India. There are important national breeds in Bulgaria, Egypt, Iran, Iraq, Italy, and Turkey. In Albania, Greece, Romania, and Yugoslavia, the river buffalo is declining. In Syria it is rare, and in Hungary nearly extinct. In Bulgaria and Italy, it is moving from a milk to a meat and milk animal. A large and expanding population is in Brazil, with two main breeds, Jafarabadi and Mediterranean; smaller populations exist in other Latin American countries and in Trinidad.

POULTRY

The global status of poultry can be assessed only in very general terms because of inadequacies in existing knowledge. Two kinds of problems prevent specific assessments. First, descriptive inventories of breeds, landraces, and types do not exist for poultry. There are a few exceptions, but overall, the range of poultry genetic resources has never been adequately measured, even at national levels. Second, national census data do not clearly distinguish among the various species of domesticated birds, and so there is not even any certain knowledge of the world population for each species. Some species are totally ignored, others may be grouped under the terms *poultry* and *waterfowl*, and sometimes the term *poultry* is used only for chickens.

There are eight species of domesticated birds that provide a significant food resource of eggs and meat. A few others also provide food for humans, but their global significance is negligible. Collectively, these can be called poultry using the broad definition that includes avian species that reproduce freely under the care of man. The eight species having significance as food producers are chickens or domestic fowl (*Gallus domesticus*), turkeys (*Meleagris gallopavo*), guinea fowl (*Numida meleagris*), Japanese quail (*Coturnix japonica*), domestic ducks (*Anas platyrhynchos*), Muscovy ducks (*Cairina moschata*), domestic geese (*Anser anser* and *Anser cygnoides*), and pigeons (*Columba livia*).

Most domesticated birds have passed through four stages in their evolution. The first was domestication per se. The second was diffusion to other countries, other environments, and other cultures. The third was the "hen craze era" of the late nineteenth and early twentieth centuries, when public interest in breeds and breeding was extremely high. And the fourth has been industrial exploitation. The result has been stratification of genetic types, some of them relic, into four categories: industrial, middle level, indigenous, and feral (Crawford, 1984; Crawford, 1990; Mason, 1984). Industrial stocks are the hybrids (two-way, three-way, and four-way crosses) bred by multinational corporations for the mass production of eggs and meat under intensive confinement conditions. Middle-level stocks are the traditional breeds of developed countries; they have reasonably good production performance if kept under reasonably good husbandry. Indigenous stocks are native to an area and are mostly kept as scavengers under conditions of minimal care and, thus, low productivity. Feral stocks have reverted to the wild and are away from the care of man.

Chickens

Chickens are the most important domestic animals globally as a source of food for man. Although they originated in the tropics, they perform well in temperate and cold climates if provided warm housing and special care. Industrial and middle-level stocks developed under temperate conditions.

Industrial chickens are bred by primary breeder operations owned by multinational corporations. Fewer than 10 primary breeders, which generate both white-shell and brown-shell layers, dominate the world industrial egg market. Fewer than 10 breeders now dominate the world industrial broiler market. There are a few small-scale primary breeder operations at regional levels.

All industrial chickens are crossbreeds, usually three-way or four-way crosses. The pure grandparent lines are kept by the primary breeder. Crossbred parent stocks are kept by multipliers and distributors who sell the commercial progeny to the producers. White-shell layers are exclusively White Leghorn strain crosses. Brown-shell layers are breed crosses, principally involving the Rhode Island Red, Barred Plymouth Rock, Black Australorp, and a few other breeds. Chicken broilers are crosses of lines based on the White Cornish and White Plymouth Rock breeds.

The major primary breeders sell their products in most developed countries, except Australia, which has its own group of primary breeders. The industrial stocks are also used extensively in centrally planned economies, either as a basic resource for local selection and reproduction or as annual replacements of production flocks, and in some developing countries that have imported high-input industrialized poultry systems.

The origin and inventory of grandparent stocks are a corporate secret of each primary breeder. It is widely suspected that many primary breeders share the same or very similar grandparent lines. Thus, the genetic base for industrial poultry may be exceedingly narrow. Most primary breeders maintain genetic resources in reserve to meet changing product requirements and production conditions. Their future breeding horizon rarely exceeds a decade, however.

Middle-level chickens are the traditional breeds developed in cold and temperate climates. Most are dual purpose, but Leghorns are egg producers. Most were developed in western Europe, Great Britain, and North America during the "hen craze" era that began in the late nineteenth century. Phenotypic descriptions can be found in fancier guidebooks, such as the *American Standard of Perfection* (1983), first published by the American Poultry Association in 1874. Several breeds, including Rhode Island Red, White Leghorn, and Barred Plymouth Rock, have been repeatedly introduced into developing countries to improve the production performance of indigenous stocks. Most indigenous stocks probably would include some genes from these middle-level breeds. These breeds were also widely used in centrally planned economies before the advent of industrial stocks.

Middle-level chickens are the most endangered poultry resource in western countries, where industrialization is most advanced. Only eight middle-level strains remained in use in Canada in 1980 (Crawford, 1984), and most are now commercially extinct. An American Minor Breeds Conservancy (1987) survey revealed a similar situation in the United States. Some middle-level breeds are being conserved by public agencies in Europe, and fanciers are active in conservation in Great Britain and North America. Except for Canada, the United States, and Australia, no satisfactory inventories of middle-level breeds exist. Particularly lacking is knowledge of these breeds in eastern Europe, Asia, and Latin America.

Indigenous chickens are found around the world in hot climates. The birds are small and hardy, but their productivity is low. They are seriously endangered globally because of the invasion by industrial stocks from Western countries and their susceptibility to pathogens introduced by the industrial stocks. Little documentation is available on their production and adaptation traits.

In Africa, only the indigenous chickens of Nigeria and Egypt have been studied in detail. Those of Central and South America are virtually unknown. The native chickens of Iran have been described, but there are no known studies from other areas of southwest Asia. Most is known about the indigenous resources of southeast Asia and Oceania (Society for the Advancement of Breeding Researches in Asia and Oceania, 1980). There are 18 indigenous chicken breeds in India, 2 each in Indonesia and Thailand, and 1 each in Japan, Malaysia, the Philippines, Taiwan, Bangladesh, and Papua New Guinea. Intermating of wild and domestic stocks is common in southeast Asia, the center of origin.

Feral populations of chickens are few in number. It is believed that the red junglefowl of the Philippines and some islands of Oceania are really feral stocks.

In addition, there are many inbred lines, random-bred control strains, stocks carrying particular blood type alleles, and lines carrying mutant genes and chromosome rearrangements. More than 350 mutations have been reported (Somes, 1988), and 125 genes from chickens have been cloned. Somes (1988) has listed 217 specialized chicken stocks.

Turkeys

Turkeys are a species of temperate climates, where they perform well if given suitable care. They do not tolerate extreme heat. They are second in importance to chickens as a meat source in Western countries. They have lesser importance in centrally planned economies. Except in Central and South America, they are seldom kept in developing countries.

Industrial turkeys are a three-way or four-way cross, always white feathered and broad breasted. They are reproduced by artificial insemination, because their natural mating ability is hampered by the broad-breasted character bred into them. Fewer than five primary breeders share most of the world market for industrial turkeys. There is no inventory of genetic resources.

Middle-level turkeys are rare. The few that remain produce less meat than industrial lines, but they can mate naturally. Only one strain remains in Canada (Crawford, 1984), and very few are left in the United States (American Minor Breeds Conservancy, 1987). A French stock is marketed, especially in Latin America.

Indigenous turkeys are prevalent in Mexico and in South America. They probably have descended directly from the original Mexican domestication without the influence of the eastern wild turkey of the United States, which contributed to development of the industrial broad-breasted turkey.

Guinea Fowl

This species is adapted to a hot, dry climate. It needs warm housing in cold climates and will not reproduce until temperatures exceed 15°C. Industrial breeding has begun in France. Guinea fowl production is a small-scale operation in western Europe and North America, but it is more important in central and eastern Europe. Guinea fowl are second to chickens in importance in most of Africa, but they are barely known elsewhere in the world.

There are no distinctive breeds. Some stocks have been selected for growth rate and body size and could be classed as middle-level breeds. Most are essentially wild types. Those of Africa could be called indigenous; eggs and meat are harvested from wild, feral, and domestic flocks.

Japanese Quail

Coturnix have gained importance as a source of eggs and meat since World War II. Those bred for food production are of the Japanese species. They are a migratory species and require a temperate or warm environment. Their use has spread rapidly to all continents, mostly for production of specialty foods. In southeast Asia, Japanese quail production is becoming a major industry.

Most stocks in use could be classed as middle-level breeds. Breeds in the traditional sense have not yet been identified, and the existence of truly indigenous stocks is not documented. Except for a partial inventory of laboratory strains and mutations (Somes, 1988), there is no known inventory of existing genetic resources.

Domestic Ducks

Mallards, the sole ancestor of domestic ducks, are a northern temperate species. Hence, the domestic form can tolerate all environments, except very hot. Their egg-production potential exceeds that of chickens, but only in southeast Asia are duck eggs a preferred product.

Industrial breeding of meat ducks is confined to developed countries. Crossbreeds are based on the White Pekin breed. Several major breeding companies provide most of the commercial ducklings.

Middle-level breeds are found in Western countries, but they are not important food sources. Khaki Campbells and Indian Runners are noted for high levels of egg production.

Indigenous stocks in southeast Asia are kept mainly for egg production under herding management. A few stocks of industrial origin are confined to meat production. These are the best-described indigenous stocks of any poultry species (Farrell and Stapleton, 1986; Society for the Advancement of Breeding Researches in Asia and Oceania, 1980). Particularly notable are the Alabio of South Kalimantan, Tegal of Java, Tsaiya of Taiwan, and Indian Runner of Indonesia and Malaysia.

Muscovy Ducks

This species is tropical and does not tolerate cold temperatures. It remains similar to the wild species, with only a few color variants and variations in body size. Muscovy ducks can be hybridized with domestic ducks (usually muscovy male × domestic female) to produce the sterile mulard, which is in demand as a meat product, especially in Taiwan.

Despite their high yield of low-fat meat, muscovy ducks have not received much attention. They are just now becoming the focus of industrial breeding in France. Most stocks in developed countries and centrally planned economies can be classed as middle level, but separate breeds and strains are not recognized. The stocks in developing countries can be regarded as indigenous. Their greatest value is in their brooding behavior and mothering ability, both toward their own progeny and those of other species.

Domestic Geese

Domestic geese are descended from two wild species. Eastern stocks are *Anser cygnoides*, and the western ones are *Anser anser*. The two types can interbreed freely. Both are temperate northern species adaptable to all but the warmest environments. Because of their poor tolerance of high temperatures, few geese are raised in tropical developing countries. They have only minor importance as a meat source in developed countries but are of greater importance in centrally planned economies, especially China.

Industrial breeding of geese has not begun. Middle-level breeds provide most of the food-producing birds, as purebreds and as crosses. The predominant breeds in Western countries are Emden, Toulouse, and Chinese. There are believed to be many middle-level stocks, especially in China, but the relationship of those stocks to the predominant breeds in Europe and North America is not known. Indigenous stocks probably exist, but they have not been described.

Pigeons

Pigeons are kept largely by fanciers in temperate and cold climates. In hot areas of Europe, Asia, and Africa, however, they have been a traditional source of meat for many centuries (Levi, 1969). The meat product is squab (young pigeon nearly ready to fly). There is a considerable but uncataloged array of genetic resources, including kinds that are highly developed for squab production (Levi, 1969).

REFERENCES

de Alba, J. 1987. Criollo cattle of Latin America. Pp. 19–44 in Animal Genetic Resources, Strategies for Improved Use and Conservation, J. Hodges, ed. FAO Animal Production and Health Paper No. 66. Rome: Food and Agriculture Organization of the United Nations.

American Minor Breeds Conservancy (AMBC). 1987. 1987 AMBC Poultry Census and Sourcebook. Pittsboro, N.C.: American Minor Breeds Conservancy.

American Poultry Association. 1983. The American Standard of Perfection, J. Skinner, ed. Troy, N.Y.: American Poultry Association.

Bowman, J. C. 1974. Conservation of rare breeds in the United Kingdom. Pp. 23–29 in Proceedings of the First World Congress on Genetics Applied to Livestock Production, Vol. II. Madrid, Spain: Graficas Orbe.

Cheng Peilieu. 1984. Livestock Breeds of China. FAO Animal Production and Health Paper No. 46. Rome, Italy: Food and Agriculture Organization of the United Nations.

Crawford, R. D. 1984. Assessment and conservation of animal genetic resources in Canada. Can. J. Anim. Sci. 64:235–251.

Crawford, R. D., ed. 1990. Poultry Breeding and Genetics. Amsterdam: Elsevier Science Publishers.

Dmitriev, N. G., and L. K. Ernst, eds. 1989. Animal Genetic Resources of the USSR. FAO Animal Production and Health Paper No. 65. Rome, Italy: Food and Agriculture Organization of the United Nations.

Farrell, D. J., and P. Stapleton, eds. 1986. Duck Production Science and World Practice. Armidale, Australia: University of New England.

Levi, W. M. 1969. The Pigeon, 2d rev. ed. Sumter, N.C.: Levi.

Maijala, K., A. V. Cherekaev, J. M. Devillard, Z. Reklewski, G. Rognoni, D. L. Simon, and D. E. Steane. 1984. Conservation of animal genetic resources in Europe.

Final report of an EAAP (European Association of Animal Production) working party. Livestock Prod. Sci. 11:3–22.
Mason, I. L., ed. 1984. Evolution of Domesticated Animals. New York: Longman.
Mason, I. L. 1988. A World Dictionary of Livestock Breeds Types and Varieties, 3d ed. Wallingford, U.K.: CAB International.
Society for the Advancement of Breeding Researches in Asia and Oceania (SABRAO). 1980. Proceedings of SABRAO Workshop on Animal Genetic Resources in Asia and Oceania. Tsukuba, Japan: Tropical Agricultural Research Center.
Somes, R. G., Jr. 1988. International Registry of Poultry Genetic Stocks. Storrs Agricultural Experiment Station Bulletin No. 476. Storrs: University of Connecticut.

APPENDIX

B

Embryo Transfer: An Assessment of the Risks of Disease Transmission

Elizabeth Singh

To date, the most extensive use of embryo transfer has been to increase the number of offspring from genetically superior females. Other uses include planned matings, genetic testing for Mendelian recessive traits, twinning (in cattle), and the salvage of desirable genetic resources from infected animals. When embryo transfer is used in conjunction with cryopreservation, it also enables the storage and international movement of genetic material. The advantages that embryos have over semen in this regard is that they provide the complete genotype.

Before techniques of embryo transfer can be used to conserve the genetic resources of all domestic species, however, additional research is required. Even if existing methods were optimal, the conservation of genetic resources by these means would only be useful if the stored germplasm were free from infectious disease. The main purpose of this appendix is to assess the disease transmission potential of embryos and to identify those factors that can influence that potential. A background on the status of embryo transfer technology for the various species is also provided. For a more detailed review, the reader should consult other articles (for example, Betteridge, 1977; Hare, 1985; Mapletoft, 1987).

Elizabeth Singh is acting director of the Animal Diseases Research Institute, Ontario, Canada.

EMBRYO TRANSFER TECHNOLOGY IN DOMESTIC SPECIES

Embryo transfer involves the collection of embryos from the genetic mother (donor) for transfer to surrogate females (recipients). Although single embryos can be collected and transferred, donors are generally superovulated to allow the collection of several embryos from each donor. A number of protocols are available for inducing superovulation in most livestock species. The gonadotropic follicle-stimulating hormone (FSH) and pregnant mare serum gonadotropin (PMSG) have both been used for this purpose; they are given during the luteal phase of the cycle, the duration of which is controlled by the administration of prostaglandins, or at the end of a period of progesterone administration.

Depending on the species, embryo collection is carried out either surgically or nonsurgically. Nonsurgical methods are preferable in that they do not damage the reproductive tract, are repeatable, and can be carried out on the farm. Generally, embryos are collected at the morula or early blastocyst stage. Using surgical methods, earlier stage embryos can be collected, although their cryopreservation is less successful. Regardless of their embryonic stage, embryos are usually collected and transferred while their zona pellucida is intact. Optimal pregnancy rates are obtained only when the preceding estrus of the donor and recipient occur at about the same time.

Prior to cryopreservation or transfer, embryos are evaluated on the basis of their morphology. Embryos rated good or excellent produce the highest pregnancy rate or survival rate following cryopreservation. The freezing process involves the slow cooling of embryos to an appropriate temperature and then direct transfer into liquid nitrogen. Thawing takes place by warming rapidly.

At present, technologies for embryo transfer have allowed for in vitro fertilization, the sexing of embryos, and the splitting of embryos into parts to produce clones. Although there has been limited use of these techniques to date, they are bound to become increasingly important. The remainder of this section is a summary of the state of embryo transfer technology for the various domestic species.

Cattle generally respond well to superovulation; they produce an average of 8 to 10 embryos/eggs per treated animal. The use of FSH would appear to be superior to the use of PMSG. Embryos are collected nonsurgically 6 to 8 days after estrus. Cryopreservation is extensively used in bovine embryo transfer programs, and transfers are now generally carried out nonsurgically. Generally, one embryo is transferred per recipient, and synchrony between the estrus cycle of the donor and the recipient is within 1 day. With existing technology, five or six of the embryos collected per superovulated cow would be of transferable quality and they would produce three or four pregnancies. Pregnancy rates are generally around 60 percent with fresh embryos and 40 to 50 percent with frozen embryos.

In sheep and goats the average number of embryos/eggs obtained with superovulation is 10. Because fertilization failure frequently occurs in superovulated ewes, semen is often deposited into the tip of each uterine horn surgically. Generally, because of the tortuous nature of the cervix, embryo collection and transfer is carried out surgically. Recently, however, nonsurgical and laparoscopic methods of collection have been used successfully. Depending on the method of collection, embryos are collected either 3 to 4 (8- to 16-cell stage) or 5 to 7 (morula to blastocyst stage) days after estrus. Both freezing and micromanipulation are successful with ovine and caprine embryos. When embryos are to be frozen, they should be collected at the late morula and early blastocyst stage. Generally, two embryos are transferred per recipient. The requirement for synchrony in the estrus of the donor and recipient is the same for sheep and goats as it is for cattle. The survival rate of good embryos is 50 to 60 percent, and the number of offspring produced per collection averages five.

In pigs synchronization can be difficult because prostaglandin is luteolytic only after day 10. Methods developed to overcome this problem include weaning piglets from sows, aborting sows 16 to 45 days after breeding, or administering prostaglandin to pseudopregnant sows. Donors can then be successfully superovulated with one injection of PMSG, but because pigs are multiple ovulators, this is unnecessary. Embryos are usually collected surgically 4 to 6 days after estrus, at which time the embryos range from the four-cell to the expanded-blastocyst stage. Although it is difficult to assess the quality of morula-stage embryos, it would appear that the best pregnancy rates are obtained with this stage or are most successful when the donor's estrus is synchronous with or occurs either 1 or 2 days prior to or 1 day after that of the recipient. On the average, 30 embryos are produced with superovulation and 20 offspring are produced per collection. Pig embryos are successfully frozen at expanded blastocyst (entire zona pellucida) and early hatched stages, but pregnancy rates are very low.

In the horse, gonadotropins are generally not useful for inducing superovulation. Equine pituitary extract has been used, but the embryos produced resulted in poor pregnancy rates. Thus, single embryos are usually collected from donors following spontaneous ovulation. For reproductively sound mares bred to a fertile stallion, embryo recovery rates can be expected to be from 50 to 80 percent. Generally, embryo collection is carried out nonsurgically 6 to 9 days after ovulation. The equine embryo loses its zona pellucida and is surrounded by a capsule about the seventh day. Transfers can be carried out surgically or nonsurgically. Pregnancy rates of about 50 percent are achieved with either method. Although equine embryos can be successfully frozen or bisected, limited use has been made of these technologies in this species.

THE DISEASE TRANSMISSION POTENTIAL OF EMBRYOS

On purely theoretical grounds, the disease transmission potential of embryos is much less than that of either the live animal or semen. Depending on the species, embryos are usually collected and transferred when they are 4 to 7 days old. Thus, prior to collection, there is a very short period of time in which an embryo can become infected. Moreover, the embryo is limited in terms of exposure to only those pathogens that are found in the reproductive tract of its mother. In addition, because embryos are collected at the zona pellucida-intact stage in most species, any pathogens in the reproductive tract must also be capable of penetrating this structure to gain access to the embryonic cells. After the seventh or eight day, equine embryos are protected by a capsule and not the zona pellucida, and thus, equine pathogens would have to be capable of crossing this protective barrier.

Other factors that help to reduce the disease transmission potential of early embryos are inherent in the techniques that are used in the collection and processing of embryos. Embryo transfer technology involves flushing embryos out of the reproductive tract with several hundred milliliters of fluid. This volume helps to dilute pathogens that might be present in the uterus. In addition, this technology allows for embryo washing and the antimicrobial and enzymatic treatment of embryos in order to enhance their freedom from disease. Finally, the majority of embryos involved in international trade are frozen, and cryopreservation has been found to be effective in inactivating low levels of the viruses that can adhere to embryos.

There are no guidelines that can be used to predict which of the disease agents might be transmitted by embryo transfer. Each pathogenic organism has to be investigated individually. In addition, because there are inherent differences in the zona pellucida of the different species of embryos, the potential for disease transmission by embryo transfer of each pathogen must be investigated in each species.

In terms of the major diseases of concern that might be transmitted by embryo transfer, however, a number of conclusions are possible. Embryos will be free of parasites, and in all likelihood, they will be protected from both bacterial and fungal agents by the presence of an intact zona pellucida. These disease agents are too large to be able to cross this structure, and even if they do adhere to it, the presence of antibiotics and antimycotics in the washing media will most likely inactivate them. Thus, the major diseases of concern will most likely be viral in nature. Each viral pathogen must, therefore, be investigated to determine its ability to penetrate the zona pellucida and infect the embryo or to adhere to the zona so firmly that it cannot be removed by washing.

MECHANISMS OF DISEASE TRANSMISSION BY EMBRYO TRANSFER

For infectious disease transmission to occur through an embryo, a disease agent has to be transferred (1) in the embryo, (2) in or on the zona pellucida of the embryo, or (3) in the fluids in which the embryo is transferred. Each of these transmission mechanisms is described below.

1. The transfer of an infected embryo. If either the oocyte or spermatozoon is infected, the resulting embryo will be infected at the time of fertilization. It is generally accepted, however, that gametic infection is not a significant factor in embryonic infection. Few pathogens have been demonstrated in oocytes, and the majority of pathogens found in semen are in the seminal fluid and not associated with the spermatozoa (Eaglesome et al., 1980). Even if agents do become adsorbed to the surface of the spermatozoa, it is unlikely that those agents would infect an embryo at fertilization because most of the outer membrane and contents of the acrosome are lost from spermatozoa that penetrate the zona pellucida. Thus, if embryos do become infected, they do so by coming in contact with a pathogen in the reproductive tract of their mother. This pathogen may have been introduced into the reproductive tract in the seminal fluid of the semen used to breed the donor or be a contaminant of the uterine excretions.

2. The transfer of an embryo with a pathogen in or on the zona pellucida. After in vitro exposure, some pathogens can adhere to the zona pellucida of embryos so strongly that washing fails to remove them (Singh, 1987). Although the embryo itself is uninfected, the transfer of an embryo carrying a pathogen on the zona pellucida can result in infection of the recipient and possibly infection of the fetus.

Thus, if a pathogen adheres to the zona pellucida under in vitro conditions, it is essential to determine whether the pathogen is ever excreted into the reproductive tract of infected animals. If it is, it might adhere to embryos in vivo and be transmitted using embryo transfer.

3. The transfer of an embryo in media contaminated with a disease agent. Recipients could become infected if embryos were transferred in contaminated media or if procedures to ensure the sanitary health of embryos were not adhered to. Proper washing (see Annex B-1) is effective in removing very high levels of infectivity from embryos as long as the pathogens do not adhere to the zona pellucida. However, the sterility of the washing, freezing, and transfer media is essential to ensuring that pathogens are not introduced and transferred along with the embryo. Products of animal origin, such as serum or bovine serum albumin (BSA), that are used in the various media may create problems. There is some evidence that bovine viral diarrhea (BVD) may have been transmitted to recipients, not from the embryos transferred but from the BSA used in the transfer media (Anderson et al., 1988). Thus, it is essential that all substances that come into contact with the embryo be sterile.

STUDIES OF THE DISEASE TRANSMISSION POTENTIAL OF EMBRYOS

Most of the research on the disease transmission potential of embryos has involved either the in vitro exposure of embryos to pathogens or the collection of embryos from acutely infected donors (in vivo experiments). The embryos are then either assayed in tissue culture or transferred to susceptible quarantined recipients.

The in vitro experiments have provided much useful information and have facilitated the development of washing and enzymatic treatments of embryos to enhance their health status. It should be remembered in assessing this work, however, that embryos are being exposed to preparations of pathogens that usually contain high levels of proteins and enzymes. These latter substances might alter the adherence of the pathogen to the zona pellucida of embryos. In addition, embryos are also being exposed to much higher levels of the pathogens in vitro than they would be under the most extreme in vivo conditions. If embryos never come into contact with a specific disease agent in vivo, demonstration of adherence under in vitro conditions is of little significance. For these and many other reasons, it is probably unwise to extrapolate from an in vitro to an in vivo situation.

The in vivo experiments have their own limitations. Generally, the best data for determining the disease transmission potential of embryos are data on the transfer of embryos from seropositive donors to quarantined susceptible recipients. Certainly, these are the donors that would most likely undergo embryo collection in the field. However, only a very few seropositive animals are actively infected and therefore capable of potentially transmitting the disease agent with their embryos. A very large number of transfers would be required to assess fully the potential of seropositive donors to transmit a particular disease by embryo transfer. These numbers would be difficult to generate at most research establishments. For this reason transfers are often carried out from actively infected (viremic) donors to maximize the possibility of disease transmission through the embryo. Since these donors have high titers of virus in their blood and other tissues, however, their embryos could become contaminated and must therefore be washed thoroughly prior to transfer.

The other variable that should be considered in assessing the research is the method by which embryo infectivity is determined. This is usually carried out by transferring in vivo- or in vitro-exposed embryos to recipients, by assaying the embryos in tissue culture, or by animal inoculation. Depending on the disease agent, the sensitivity of each of these detection methods will vary. It is generally accepted, however, that tissue culture systems are more sensitive for the detection of virus than are animals when only one embryo is transferred per utero. In fact, tissue culture assays have been sensitive enough to detect the small number of virions that adhere to the zona pellucida of a single embryo (Singh and Thomas, 1987b; Singh et al., 1982b, 1984, 1987). Thus, before a final conclusion can be reached on the transmissibility of each disease agent, the assay system used to generate the results must be assessed in terms of its sensitivity and specificity.

TRANSMISSIBILITY OF SPECIFIC AGENTS BY EMBRYO TRANSFER

The following is a comprehensive review of the work that has been carried out on the disease transmission potential of embryos. Much of the work was undertaken before development of the procedures for washing and processing embryos that are recommended by the International Embryo Transfer Society and endorsed by the Office International des Épizooties (see Annex B-1). Thus, there is considerable variation in the studies in how embryos were handled after exposure and prior to being assayed. Unless otherwise indicated, however, the methodology used in each study was shown to be effective under the conditions used. Similarly, unless otherwise indicated, all embryos exposed to pathogens in the studies were zona pellucida-intact embryos. For ease of reference, the disease agents that have been investigated are listed in alphabetical order and not in order of importance. Tables summarizing the research data are presented in Annex B-2.

The conclusions reached regarding the transmissibility of each disease agent by embryo transfer are based on the data available. One of the most difficult tasks remains—to determine the validity of extending the conclusions derived from these data to field conditions.

African Swine Fever Virus

Both in vitro (Singh et al., 1984) and in vivo experiments have been carried out with this virus. Porcine embryos were exposed to African swine fever virus (ASFV), washed, and assayed in tissue culture, and embryos were collected from viremic pigs, washed, and assayed in vitro.

Results and Conclusions

1. In vitro: Ninety-five percent of the embryos retained infectious ASFV after viral exposure and washing. Treating the embryos with papain, EDTA, or ficin had no effect on the retained virus, whereas treating them with trypsin or pronase reduced the number of embryos carrying detectable virus (30 percent instead of 95 percent) and lowered the amount of virus on the embryos. The evidence suggested that most, if not all, of the virus was on the zona pellucida. The data indicated that if ASFV is excreted into the reproductive tract of infected animals, the virus could be transmitted by embryo transfer.

2. In vivo: A total of 245 porcine embryos were collected from viremic donors, washed, and assayed in vitro in groups of 18 to 20 in order to duplicate embryo transfer conditions. None of the embryo samples was found to be associated with ASFV (Dulac and Singh, unpublished data), and very little virus was isolated from the uterine flush fluids. Thus, these preliminary results suggest that African swine fever can be controlled using embryo transfer.

Akabane Virus

In vitro experiments have been carried out with the Akabane virus (AV) (Singh et al., 1982a). Bovine embryos were exposed to AV, washed, and then either assayed or cultured and assayed.

Results and Conclusions

Akabane virus did not infect zona pellucida-intact bovine embryos, and it had no effect on in vitro embryonic development. Proper washing was effective in rendering Akabane-exposed embryos free of this virus.

Bluetongue Virus

Experiments in cattle (Acree, 1988; Bowen et al., 1982, 1983; Singh et al., 1982a; Thomas et al., 1983, 1985), sheep (Gilbert et al., 1987; Hare et al., 1988), and goats (Chemineau et al., 1986) have been carried out with bluetongue virus (BTV).

The bovine studies involved both in vitro and in vivo experiments. Both zona pellucida-intact and zona pellucida-free bovine embryos were exposed to BTV, washed, and either assayed directly or cultured and then assayed. In addition, embryos were collected from bluetongue viremic donors bred with uninfected semen and from uninfected donors bred with BTV-infected semen. These embryos were washed and then transferred to susceptible quarantined recipients.

The ovine experiments involved the transfer of embryos from bluetongue viremic sheep bred by either infected or uninfected rams to quarantined susceptible recipients. In addition, embryos were exposed to BTV, washed, and transferred to susceptible recipients.

The caprine work involved the transfer of embryos from BTV-seropositive herds to uninfected recipients.

Results and Conclusions

1. BTV did not infect zona pellucida-intact bovine embryos after in vitro exposure, nor did it have any effect on in vitro embryonic development. Thus, proper washing was effective in rendering BTV-exposed bovine embryos free of this virus. The integrity of the zona pellucida is important because BTV can infect zona-free bovine embryos. Bluetongue was not transmitted by embryo transfer. A total of 334 embryos, collected from infected donors or uninfected donors bred with infected semen, were washed and transferred into 330 quarantined susceptible recipients. Although BTV virus was isolated from some of the uterine flush fluids, none of the calves or recipients developed BTV antibodies. Thus, embryo transfer was shown to be effective in preventing the spread of bluetongue in cattle.

2. The results obtained when BTV-exposed ovine embryos were

transferred to uninfected recipients varied. In one study 49 embryos were transferred into 27 recipients, and all of the lambs and recipients remained BTV seronegative. In the other study, when 20 embryos from BTV-infected sheep were transferred into 15 recipients, 2 of the recipients seroconverted, and when 13 embryos exposed to BTV virus in vitro were transferred, 9 of the 13 recipients seroconverted. The difference in the results might be due to the different strains of virus used in the two studies. In the study in which transmission occurred, however, the embryos were washed only 4 times instead of the recommended 10. Since the investigators did not establish that the four washings were effective, implication of the ovine embryo in the transmission of bluetongue is uncertain. Additional research is required to determine whether BTV can be transmitted to sheep using embryo transfer when the embryos are handled according to the recommendations of the International Embryo Transfer Society.

3. When 63 caprine embryos derived from herds in which 47 percent were BTV seropositive were transferred to 19 recipients, the recipients and kids remained BTV seronegative. These initial experiments indicate that embryo transfer may be successful in preventing the transmission of BTV.

Bovine Leukemia Virus

Both in vitro (Hare et al., 1985) and in vivo (Di Giacomo et al., 1986; Eaglesome et al., 1982; Hare et al., 1985; Kaja et al., 1984; Olson et al., 1982; Parodi et al., 1983; Thibier and Nibart, 1987) experiments have been carried out with bovine leukemia virus (BLV). Zona pellucida-intact and zona-free bovine embryos have been exposed to the virus in vitro, washed, and then either assayed or transferred to uninfected recipients. Embryos also have been transferred from BLV-seropositive donors to uninfected recipients.

Results and Conclusions

1. When 27 zona-intact and 15 zona-free bovine embryos were exposed to BLV, washed, and assayed, no infectivity was associated with any of the embryos. Similarly, when 48 embryos were exposed to the virus, washed, and transferred into 3 recipients, the recipients remained BLV seronegative. These experiments indicate that proper washing is effective in rendering embryos free of BLV.

2. At least 1,200 embryos (596 from known BLV-seropositive donors and the remainder from 1,500 donors in which the majority were BLV seropositive) have been transferred from BLV seropositive do-

nors to uninfected recipients, and all of the recipients and calves remained BLV seronegative. This evidence demonstrates that BLV is highly unlikely to be transmitted by embryo transfer.

Bovine Viral Diarrhea Virus

Both in vitro (Evermann et al., 1981; Potter et al., 1984; Singh et al., 1982b) and in vivo (Archbald et al., 1979) experiments have been carried out with bovine viral diarrhea virus (BVDV). Bovine embryos were exposed to BVDV in vitro, washed, and assayed using two assay systems in order to maximize the chances of detecting virus in or on the embryo. Both zona-intact and zona-free ovine embryos were exposed to BVDV in vitro, washed, and then assayed.

Bovine embryos were collected from donors in which BVDV was inoculated into one of their uterine horns. These embryos were assessed and then examined using electron microscopy.

Results and Conclusions

1. BVDV did not infect zona pellucida-intact bovine or ovine embryos, nor did it have any effect on in vitro embryonic development. Proper washing was effective in removing BVDV from all zona pellucida-intact embryos. This virus also did not replicate in zona-free ovine embryos.

2. The results from eight embryos obtained from donors inoculated per utero with BVDV are difficult to interpret. The embryos were degenerating and there was evidence of structures that morphologically resembled BVDV beneath the zona pellucida. The degeneration was most likely caused by inflammation in the uterine horn. The significance of the particles will require further investigation, however. Because the embryos were degenerating, the zona pellucida might not have been an effective barrier to the virus, which might account for the BVDV-like particles beneath the zona.

Brucella abortus

Both in vitro (Mallek et al., 1984; Stringfellow et al., 1984) and in vivo (Barrios et al., 1988; Bolin et al., 1981; Stringfellow et al., 1982, 1983, 1988; Voelkel et al., 1983) experiments have been carried out with this agent. Bovine embryos have been exposed to the agent, washed, and assayed, and embryos have been transferred from seropositive donors to susceptible recipients.

Results and Conclusions

Brucella abortus was not isolated from any of 96 embryos that were exposed to this agent and washed. Similarly, 39 embryos transferred from *B. abortus*-seropositive donors to uninfected recipients did not result in seroconversion of the recipients, and 309 embryos from infected donors were not associated with this agent when assayed. Of 116 uterine flush fluids from *B. abortus*-infected donors, the organism was isolated from only 9. This evidence indicates that *B. abortus* is unlikely to be transmitted through embryo transfer.

Caprine Arthritis and Encephalitis Virus

Some in vivo work has been carried out with caprine arthritis and encephalitis virus (CAEV) (Wolfe et al., 1987). Embryos from CAEV-seropositive donors were transferred to uninfected recipients.

Results and Conclusions

When 16 embryos were collected from CAEV-seropositive donors and transferred to susceptible recipients, the recipients remained CAEV seronegative. Additional transfers are required to support these preliminary results indicating that CAEV is not transferred using embryo transfer.

Foot-and-Mouth Disease Virus

The transmission of foot-and-mouth disease virus (FMDV) has been studied in both cattle (McVicar et al., 1986; Mebus and Singh, 1988; Singh et al., 1986) and pigs (Mebus and Singh, 1988; Singh et al., 1986), and both in vitro and in vivo studies have been carried out. Bovine and porcine zona pellucida-intact embryos and bovine zona-free embryos were exposed to high levels of infectivity of FMDV, washed, and assayed in tissue culture and steer tongue. Both bovine and porcine embryos were collected from FMDV-viremic donors. Some of the embryos were transferred, and the balance were assayed in tissue culture or steer tongue.

Results and Conclusions

1. No infectious virus was detected on any of the 169 bovine zona pellucida-intact embryos exposed in vitro to FMDV, washed,

and assayed. Thus, proper washing was effective in rendering bovine embryos free of FMDV. This was not true for porcine embryos. Five percent of 194 zona pellucida-intact porcine embryos were associated with FMDV after exposure and washing. Incubation of the embryos resulted in a decrease in the amount of virus detected, which indicates that FMDV was not replicating in the embryonic cells of the porcine embryos and that the virus was most likely on the zona pellucida.

2. Once the zona pellucida was removed from bovine embryos, proper washing was no longer effective in removing this virus from the embryos. Thirty percent of bovine zona-free embryos carried FMDV after in vitro exposure and washing. Infectivity titers were higher in hatched embryos after culture than in those assayed immediately, although there was no increase in titer in embryos incubated for 2 days compared with those incubated for 1 day. The virus may have entered the embryonic cells, although it is also possible that the virus had merely become trapped in the folds of the embryo.

3. One hundred forty-nine bovine zona pellucida-intact embryos were collected from FMDV-viremic donors, washed, and transferred into 112 quarantined susceptible recipients. Fifteen of the uterine flush fluids from 22 of the donors were found to contain small amounts of FMDV. All calves and recipients have remained FMDV seronegative. In addition, 372 embryos and 64 eggs collected from these donors were assayed in steer tongues or tissue culture and found to be negative for infectivity. This evidence indicates that the transmission of FMDV can be controlled by embryo transfer.

4. FMDV was not isolated from any of the 177 embryos and 104 eggs collected from FMDV-viremic pigs, washed, and assayed in steer tongue. In order to mimic embryo transfer conditions, embryos were assayed as group samples (20 embryos per sample). Four of the fifteen uterine flush fluid samples from the infected donors were found to be contaminated with low levels of FMDV. In addition, 436 embryos and eggs were collected from viremic pigs, washed, and transferred into 13 quarantined recipients. All of the recipients remained FMDV seronegative. This evidence indicates that embryo transfer is useful in controlling the transmission of foot-and-mouth disease in pigs.

Haemophilus somnus

Some in vitro experiments have been carried out with this agent (Thomson et al., 1987). Bovine embryos were exposed to *Haemophilus somnus* and then washed in the presence or absence of antibiotics.

Results and Conclusions

Haemophilus somnus was isolated from 10 of 38 embryos after washing in the absence of antibiotics and from none of 9 embryos after washing in the presence of antibiotics. The results show that antibiotics must be included in the flushing and washing media to prevent the multiplication of the organisms that adhere to the zona pellucida of embryos.

Hog Cholera Virus

Both in vitro (Dulac and Singh, 1987) and in vivo experiments have been carried out with this disease agent. Embryos were exposed to hog cholera virus (HCV) in vitro, washed, and assayed. In addition, embryos were collected from HCV-viremic donors, washed, and transferred to uninfected quarantined recipients.

Results and Conclusions

1. HCV was found to adhere to the zona pellucida when 171 porcine embryos were exposed to the virus and washed. This virus had no effect on the in vitro development of the embryos, and trypsin treatment was found to be effective in rendering embryos noninfectious if they had been exposed to less than 10^6 infectivity units of HCV. If the virus is excreted into the reproductive tract of infected animals, trypsin treatment would be required to render the embryos noninfectious.
2. Three hundred and eighty-five porcine embryos were collected from HCV-viremic donors, washed, and transferred to uninfected quarantined recipients. Nineteen embryos and seventy-four unfertilized eggs were also assayed in vitro. The recipients remained HCV seronegative and the embryos/eggs were all found to be negative for HC infectivity. These preliminary results indicate that embryo transfer could be used to control the transmission of hog cholera.

Infectious Bovine Rhinotracheitis Virus

Both in vitro (Bowen et al., 1985; Singh et al., 1982b) and in vivo (Bondioli et al., 1988; Echternkamp and Maurer, 1988; Hasler and Reinders, 1988; Singh et al., 1982b, 1983) studies of infectious bovine rhinotracheitis virus (IBRV) have been carried out. (1) Various stages of bovine embryos were exposed for 1 or 24 hours to IBRV in vitro, washed, and either assayed or cultured and assayed. In addition,

IBRV-exposed embryos were washed, trypsin treated, and assayed. (2) Embryos were collected from IBRV-viremic donors, washed, trypsin treated, and transferred to susceptible quarantined recipients. (3) Embryos were exposed to IBRV in vitro and transferred to susceptible recipients after being washed or washed and trypsin treated. (4) Embryos were collected from uninfected donors bred with IBRV-infected semen, washed, and assayed in tissue culture. (5) The effect of treating embryos with trypsin both on the pregnancy rate and the ability of embryos to survive cryopreservation was determined.

Results and Conclusions

1. IBRV was isolated from 60 percent of virus-exposed embryos, although it had no effect on the in vitro embryonic development of the embryos. Trypsin (0.25 percent, 60 seconds) and IBRV antiserum were found to be capable of rendering the IBRV-exposed embryos noninfectious. Both the low level of the virus isolated from the embryos and the susceptibility of this virus to trypsin and antiserum suggest that IBRV attaches to the zona pellucida of embryos and cannot penetrate it to gain access to the embryonic cells. Once the zona is removed, IBRV replicates in the embryonic cells. If very high levels of the virus are excreted into the reproductive tract of infected animals, the data indicate that trypsin treatment would be necessary to render the embryos noninfective.

2. Sixty-four embryos collected from twenty-two donors infected with IBRV were transferred after washing and trypsin treatment to forty-nine uninfected quarantined recipients. Twenty-three pregnancies resulted from these transfers, and all of the recipients and resulting calves remained IBRV seronegative. Thus, embryos can be transferred from IBRV-infected donors (viremic) without transmitting the disease if the embryos are trypsin treated prior to transfer. Since all embryos were treated, it could not be determined whether this treatment is essential in order to prevent disease transmission.

3. Thirty-eight embryos were exposed to 10^7 infectivity units of IBRV, treated with trypsin, and transferred to twenty-two uninfected recipients. All of the recipients and resulting calves remained IBRV seronegative. When eight IBRV-exposed and washed embryos were transferred without trypsin treatment into four recipients, one of the four recipients seroconverted (Singh, unpublished data). Thus, when embryos are exposed to high levels of IBRV in vitro, trypsin treatment is required to prevent disease transmission.

4. Forty-seven embryos were collected from six donors bred with IBRV-infected semen. The embryos were washed ten times and then

assayed in tissue culture. Twelve embryos (from four donors) were found to be associated with small amounts of IBRV (Singh, unpublished data). This experiment indicates that if IBRV-infected semen were used to breed donors for embryo transfer, the embryos collected would require trypsin treatment to render them noninfectious.

5. A number of studies have examined the effect of trypsin treatment on embryo viability (Bondioli et al., 1988; Echternkamp and Maurer, 1988; Hasler and Reinders, 1988). Following trypsin treatment, 80 embryos were transferred fresh and 416 were transferred after cryopreservation and thawing. The data demonstrated that trypsin treatment had no effect on the embryo's ability to survive freezing or on the pregnancy rate obtained with treated embryos.

Porcine Parvovirus

In vitro experiments have been carried out with porcine parvovirus (PPV) (Wrathall and Mengeling, 1979a,b). Embryos were exposed to the virus in vitro, washed, and then either assayed or transferred to uninfected recipients.

Results and Conclusions

When 38 porcine embryos were exposed to PPV in vitro, washed, and assayed, the virus was found to adhere to the zona pellucida of all of the embryos. When 76 embryos were exposed to the virus in vitro, washed, and transferred into four uninfected recipients, the recipients became PPV seropositive. These data would indicate that if this virus is excreted into the reproductive tract, embryo transfer might result in the transmission of this disease.

Pseudorabies Virus

Both in vitro (Bolin et al., 1981, 1982) and in vivo (Bolin et al., 1982; James et al., 1983) studies of pseudorabies virus (PrV) have been carried out in swine. (1) Embryos were exposed to PrV in vitro, washed, and assayed for infectivity. (2) Embryos were exposed to high levels of infectivity of PrV in vitro, washed, trypsin treated, and then assayed. (3) Embryos were exposed to PrV in vitro, washed, or washed and trypsin treated, and then transferred to uninfected recipients. (4) Embryos were transferred from PrV-infected donors to uninfected recipients. (5) Embryos were collected from PrV-seropositive donors and transferred to uninfected recipients.

Results and Conclusions

1. In one study 155 embryos exposed to 10^4 to 10^8 infectivity units of PrV were not associated with infectivity after washing. All embryos were cultured for 24 to 48 hours prior to assay, however, which may have resulted in some loss of viral infectivity. In another study 127 embryos were exposed to similar levels of the virus, washed, and then assayed. All embryos were negative if the exposure level was below 10^6 infectivity units and positive if above 10^6. These results suggest that at high levels of infectivity, PrV can adhere to the zona pellucida and that washing is not effective in rendering embryos noninfectious.

2. When 45 embryos were exposed to 10^6 infectivity units of PrV, washed, and trypsin treated, none of the embryos was associated with infectivity. Thus, trypsin is effective in removing PrV that adheres to the zona pellucida of embryos.

3. When 45 embryos were exposed to 10^4 infectivity units of PrV, washed, and transferred into 4 uninfected recipients, none of the recipients seroconverted; when 79 embryos were exposed to 10^8 units of the virus, washed, and transferred, all 5 of the recipients seroconverted. These results confirm earlier work indicating that washing is not effective in rendering embryos noninfectious when they have been exposed to high levels of PrV. In another study (Singh and Thomas, unpublished data) embryos were exposed to 10^7 units of PrV, washed or washed and trypsin treated, and then transferred to uninfected recipients. The 10 pigs that received the trypsin-treated embryos remained PrV seronegative, but the 6 recipients that received the washed embryos became PrV seropositive. Thus, trypsin was shown to be effective in rendering in vitro-exposed embryos noninfectious.

4. When 45 embryos were collected from donors infected intranasally with PrV, washed, and transferred into 3 uninfected recipients, none of the recipients seroconverted. When 70 embryos were collected from donors that had been infected intrauterinely and intranasally, washed, and transferred, 2 of 5 recipients became PrV seropositive. Again, these results testify to the fact that if embryos come into contact with high levels of PrV, the virus adheres to the embryos.

5. A total of 805 embryos from 38 PrV-seropositive donors have been collected and transferred to 34 uninfected recipients to produce 208 piglets. The recipients and all of the piglets remained uninfected. These results indicate that PrV-seropositive donors do not excrete significant amounts of virus into their reproductive tracts and that embryo transfer can be used for the control of PrV.

Rinderpest Virus

Both in vitro and in vivo experiments (Mebus and Singh, 1988) have been carried out with rinderpest virus (RPV). A total of 61 zona pellucida-intact bovine embryos were exposed to high levels of infectivity of RPV in vitro, washed, and assayed in tissue culture. Embryos were also collected from RPV-infected donors, washed, and then assayed by animal inoculation, in tissue culture, or by transfer to recipients.

Results and Conclusions

1. Preliminary results indicate that 3 percent of zona pellucida-intact bovine embryos are associated with RPV after in vitro exposure and washing (Dulac and Singh, unpublished results).

2. None of 170 zona pellucida-intact bovine embryos collected from RPV-viremic donors was associated with this virus after washing. The embryos were assayed by animal inoculation and in tissue culture. In addition, 17 embryos from RPV-viremic donors were transferred into 15 recipients. All of them remained RPV seronegative (Singh and Mebus, unpublished data). These data indicate that rinderpest is not transmitted through the embryo.

Scrapie

In vivo experiments have been carried out with this agent (Foote et al., 1986). Embryos were transferred from scrapie-infected sheep to uninfected recipients.

Results and Conclusions

The transfers from scrapie-infected recipients have resulted in 69 lambs. No scrapie has been identified in these animals, who at this writing now range in age from 41 to 82 months. Scrapie was diagnosed in the positive controls at approximately 40 months. Although this evidence is promising, additional studies are required before these results can be extrapolated to naturally infected animals.

Swine Vesicular Disease Virus

Both in vitro (Singh et al., 1987a) and in vivo (Singh et al., 1987b) experiments have been carried out with swine vesicular disease virus

(SVDV). In one study embryos were exposed to SVDV, washed, and assayed. In the other study embryos were collected from viremic donors and transferred to uninfected quarantined recipients.

Results and Conclusions

1. Infectious SVDV was isolated from all of the embryos after in vitro exposure and washing. Culturing the embryos for 24 or 48 hours or treating the embryos with pronase, trypsin, or antiserum after viral exposure and washing reduced the number of embryos carrying the virus and lessened the amount of virus on each of the embryos. None of the treatments, however, was capable of decontaminating all of the embryos. Thus, if SVDV is excreted into the reproductive tract of infected animals, transmission of this disease by embryo transfer is possible.

2. Two hundred and five embryos were collected from thirteen viremic donors and transferred to nine uninfected quarantined recipients. Seventeen embryos and ninety-five unfertilized eggs were also tested in vitro. The recipients, piglets, and embryos were all negative for SVDV infectivity, which indicates that control of this disease through embryo transfer is possible.

Vesicular Stomatitis Virus

Experiments have been carried out with both bovine and porcine zona pellucida-intact embryos to determine their in vitro susceptibility to vesicular stomatitis virus (VSV) (Lauerman et al., 1986; Singh et al., 1987). The embryos were monitored for both infection and the effect, if any, the virus had on embryonic development.

Results and Conclusions

VSV was found to adhere to the zona pellucida of both bovine and porcine embryos, although it had no effect on the in vitro embryonic development of these embryos. Trypsin treatment was found to be effective in rendering both porcine and bovine embryos noninfective for VSV. The need for treating embryos derived from VSV-infected donors with trypsin remains unknown. If infected donors do excrete high levels of this virus into their reproductive tract, these data would indicate that trypsin treatment would be necessary to render the embryos noninfectious.

CONCLUSIONS

An analysis of the data supports a number of conclusions regarding the disease transmission potential of embryos. Of all the disease agents that were tested, none replicated in the embryonic cells of zona pellucida-intact embryos. It was shown that some viruses (infectious bovine rhinotracheitis virus and bluetongue virus) could replicate in the embryonic cells of zona pellucida-free embryos, and thus, the integrity of the zona is an important factor in the health status of embryos. In addition, proper washing in the presence of antibiotics was shown to be effective in rendering zona pellucida-intact bovine embryos free of many disease agents (Akabane virus, bluetongue virus, bovine leukemia virus, bovine viral diarrhea virus, and foot-and-mouth disease virus, as well as *Brucella abortus* and *Haemophilus somnus*).

Two, and possibly three, bovine viruses (infectious bovine rhinotracheitis virus, vesicular stomatitis virus, and possibly rinderpest virus) and all of the porcine viruses tested were found to adhere to the zona pellucida of embryos after in vitro viral exposure and washing. If these agents can also adhere to embryos under in vivo conditions, transmission of the diseases by embryo transfer will depend on whether these agents are ever excreted into the reproductive tract of infected animals. Trypsin treatment was effective in removing or inactivating some of these viruses, and the interpretation can be made that, generally, enveloped viruses (hog cholera virus, infectious bovine rhinotracheitis virus, pseudorabies virus, and vesicular stomatitis virus) are susceptible to trypsin treatment, whereas nonenveloped viruses are not. There would appear to be a limit to the effectiveness of using trypsin to render embryos noninfectious. Embryos exposed to high levels of certain viruses (higher than 10^6) in vitro are not rendered completely noninfectious with trypsin treatment (Dulac and Singh, 1987). This level of pathogenicity, however, greatly exceeds that which might be found in the reproductive tract of infected animals to which embryos would be exposed in vivo.

None of the bovine viruses was transmitted when embryos were transferred from infected donors (bluetongue virus, bovine leukemia virus, foot-and-mouth disease virus, or infectious bovine rhinotracheitis virus) to susceptible recipients. All embryos transferred from IBRV-viremic donors were trypsin treated prior to transfer. The necessity of carrying out this procedure, however, has not yet been clearly established. Certainly, embryos collected from donors bred with IBRV-infected semen or embryos exposed in vitro to IBRV require trypsin treatment to render them noninfectious. However, large numbers of embryos have been transferred out of IBRV-seropositive donors into uninfected recipients without disease transmission. Thus, the requirement of many importing countries that embryos derived from IBRV-seropositive donors be trypsin treated warrants further investigation. Recent evidence shows that if trypsin treatment is necessary, it does not interfere with embryo viability or freezability (Bondioli et al., 1988; Echternkamp and Maurer, 1988; Hasler and Reinders, 1988).

Although some porcine viruses (African swine fever, hog cholera, and swine vesicular disease) adhered to embryos under in vitro conditions, embryos collected from donors infected with those viruses were noninfectious when assayed in groups of 15 to 20. In addition, embryos from infected donors (HCV or SVDV) did not transmit the diseases when they were transferred to uninfected recipients. Since the viruses were found in the uterine flush fluids of infected animals, it is possible that the total amount of virus that adhered to 15 to 20 embryos from infected pigs was below the minimum level detectable by both the recipient animals and the in vitro assay system. This supposition is very unlikely to be valid, however, because the in vitro assay systems have been shown to be sensitive enough to detect the small amount of virus that adheres to the zona pellucida of a single embryo (Singh et al., 1982a, 1984, 1986, 1987). The other possibility is that the binding or adherence of viruses to embryos is an in vitro artifact. Viral preparations grown and harvested in vitro contain a number of cellular components which might increase the affinity of viruses for embryos.

Only one disease agent (pseudorabies virus) is known to have been transmitted when porcine embryos were transferred from infected donors to susceptible recipients. The donor pigs were infected by introducing a large amount of virus into the uterus. Disease transmission did not occur when donors were infected by the intranasal route. Since it has been demonstrated that PrV adheres to the zona pellucida of pig embryos, these results are not surprising. However, since over 800 embryos have been transferred out of PrV-seropositive donors without any disease transmission, it would appear that under normal circumstances this virus is not excreted in significant amounts into the reproductive tract.

Fewer transfers have been carried out from infected pigs to susceptible pigs than from infected cattle to susceptible cattle; consequently, a final conclusion on the safety of using embryo transfer for disease control in the pig is not possible. It is important to remember, however, that fewer transfers will be required in the pig because the chances of detecting any disease transmission via embryo transfer are much greater: 15 to 20 porcine embryos are transferred per recipient in contrast to 1 per recipient in cattle.

Limited work has been carried out on the disease transmission potential of sheep and goat embryos. Preliminary evidence would indicate that it may be possible to control scrapie using embryo transfer. There is also no final conclusion regarding the feasibility of using this technique to control bluetongue in sheep. In one study disease transmission did not occur. In the other study bluetongue was transmitted when embryos were transferred from bluetongue-infected donors or exposed to the virus in vitro and then transferred to susceptible recipients. Unfortunately, the embryos were washed a maximum of four times in the study in which disease transmission occurred. This is far below the 10 washes recommended by the International Embryo Transfer Society, which have been proven to be effective. Since low levels of many pathogens have been isolated from the uterine flush fluids of infected donors, proper washing of embryos is essential. Before a final conclusion can be reached regarding the transmissibility of BTV through the ovine embryo, further experimentation is required. Finally, goat embryos have been transferred from herds infected with bluetongue and caprine arthritis encephalitis to uninfected herds with no resulting disease transmission.

CONSTRAINTS TO THE INTERNATIONAL EXCHANGE OF EMBRYOS

At present, there is no technology for determining the health status of embryos without destroying them. Thus, until very recently, regulations for the importation of embryos into most countries related to the donor and sire and to the herd of origin rather than to the embryo itself. Recently, based on the research that has been carried out on the disease transmission potential of embryos, the Office International des Épizooties has modified its recommendations. The recommendations recognize that there are two main ways to ensure that embryos are free of pathogenic organisms.

The traditional method is based on the testing of donor animals over extended periods of time. The health of a donor is usually assessed on the basis of the animal's serological status, not on the absence or presence of disease agents in the donor. Problems can arise with this method of assessment, however. Animals can be persistently infected with some disease agents and yet remain serologically negative for those agents (immunotolerant). Immunotolerance has been conferred on animals that are congenitally infected with bovine viral diarrhea/mucosal disease, border disease, or bluetongue. The converse of this situation also exists. Many donors have been exposed to disease agents and become serologically positive to them.

Those donors are, however, no longer infected with the disease agent, although their embryos would remain ineligible for export.

The other method is based on research on the transmissibility of agents by embryo transfer and requires that embryos be processed in accordance with the procedures of the International Embryo Transfer Society. It obviates long periods of isolation and repeated testing of donor animals. This new method is based on the proper collecting, processing, and transfer of embryos (see Annex B-1). Although the protocols have been demonstrated to be reliable for a number of disease agents, the health status of the embryo is entirely dependent on the care taken during the embryo collection and processing procedures. For the more serious diseases, many countries might insist on having their own people involved in all stages of the flushing, collection, washing, and freezing to monitor quality control. In addition, the recipients receiving the embryos might be quarantined in a maximum security quarantine station.

Some countries, in an effort to monitor the health status of imported embryos, also require the testing of the uterine flush fluid, the embryo washes, or unfertilized eggs/degenerating embryos from the same collection. The testing of these samples is more complex than carrying out serological tests on donors. Serum samples are stable and many laboratories have experience in performing such tests. The isolation of a pathogen from the uterine flush fluid or embryos/eggs will depend not only on the assay system used but also on the stability of the agent, the treatment of the sample prior to testing, and the volume tested.

RECOMMENDATIONS

Additional research is required to improve the basic techniques involved in embryo transfer. Better and more reliable methods of superovulation, embryo assessment, and cryopreservation would be of enormous benefit. In addition, improvements are also possible in the techniques used to collect and transfer embryos. Finally, research is required in related areas of embryo transfer, such as embryo splitting, cloning, sexing, and in vitro fertilization. If these techniques are to be developed to their full potential, this research must be supported. Because this appendix is primarily concerned with the potential of embryos to transmit disease, the specific recommendations for research that follow are confined to that area.

- Additional research is required to enable a conclusion on the transmissibility of all disease agents of concern through embryo transfer.

At this time a final conclusion on the transmissibility of many pathogens by embryo transfer is not possible. The evidence does strongly suggest that if disease transmission ever occurs through the embryo, it is a relatively rare event. Additional studies are required, however, especially in the nonbovine species. Since swine embryos do not survive cryopreservation, extensive research into the disease transmission potential of swine embryos may be premature. For the other species, research should be directed in terms of disease agents by the disease status of the country of origin of the desired genetic material. It should be recognized, however, that research facilities will never be able to generate sufficient data to prove unequivocally that infectious disease agents cannot be transmitted by embryo transfer. In the interim, current data can be substantiated by transferring embryos into recipients that are held in quarantine. This procedure would allow for the generation of valuable information without risking the spread of disease.

- The disease agents that can be excreted into the reproductive tract of infected animals must be identified. Some disease agents adhere to the zona pellucida of embryos in vitro, and yet embryos collected from donors infected with those agents do not transmit disease when they are transferred to susceptible recipients. The most probable reason for this is that the disease agents are not excreted into the reproductive tracts of infected animals and, therefore, do not come into contact with the embryo. For this reason it is extremely important to determine whether pathogens that adhere to the zona pellucida in vitro are ever excreted into the reproductive tract.

- The effectiveness and the nature of the antibiotics in both the flushing and washing media must be determined so that embryos will be free of bacterial agents. Very few bacterial or fungal agents have been investigated in regard to their potential for transmission by embryo transfer. The reason for this is that these disease agents are too large to be able to penetrate the zona pellucida to gain access to the embryo. However, bacterial and fungal agents can adhere to the zona pellucida. Thus, it is very important that in vitro studies be carried out to determine the efficacy of antibiotics in the collection and washing media as a means of preventing the transmission of bacterial diseases through the embryo. It is essential that the proper antibiotic be used and that sufficient time be allowed for it to act.

- Treatments must be identified that can be applied to embryos to render them free of all infectious agents without hindering embryonic viability. Potentially, this area of research could provide the greatest benefit. The nature of the interaction between certain disease agents and the zona pellucida should be studied with a view to identifying the factors that influence this interaction. With this knowledge it might be possible to expose all collected embryos, prior to transfer, to conditions that would render them disease free. Although trypsin has been used in this regard to a limited extent, it is not the treatment of choice. Trypsin is a biological product that has a potential for contamination itself, and it would appear to be most effective in inactivating enveloped viruses. Thus, it is essential to identify other agents that can be used to render embryos disease free.

- Research is required into the suitability of using synthetic media in the collection, cryopreservation, and transfer of embryos. To ensure that embryos are not contaminated during collection or processing, it is essential that serum or any product of animal origin used in the media be free of infectious agents. This requirement can be difficult to meet because serum and BSA fraction V can be contaminated with BVDV and other viruses. Rigorous testing is required to identify the presence of these agents, which are often missed on initial screening. Thus, if a synthetic medium is developed that enables the successful collection, freezing, and transfer of embryos, it would be an important gain in ensuring the disease-free status of embryos.

REFERENCES

Acree, J. A. 1988. Report to the U.S. Animal Health Association's Committee on bluetongue and bovine retrovirus. P. 124 in Proceedings of the 92d Annual Meeting of the U.S. Animal Health Association. Richmond, Va.: U.S. Animal Health Association.

Anderson, J. B., H. Pedersen, and L. Ronsholt. 1988. A series of calves born after embryo transfer and persistently infected with BVD-virus. P. 29 in Proceedings of the 13th Conference of the OIE Regional Commission for Europe. Paris, France: Office International des Épizooties.

Archbald, L. F., R. W. Fulton, C. L. Seager, F. Al-Bagdadi, and R. A. Godke. 1979. Effect of bovine viral diarrhea (BVD) virus on preimplantation bovine embryos: A preliminary study. Theriogenology 11:81–89.

Barrios, D. R., D.C. Kraemer, E. Bessoudo, and L. Adams. 1988. Failure to isolate *Brucella abortus* from embryos or ova from culture-positive superovulated cows. Theriogenology 29:353–361.

Betteridge, K. J. 1977. Embryo Transfer in Farm Animals: A Review of Techniques and Applications. Agriculture Canada Monograph 16. Ottawa, Ontario: Agriculture Canada.

Bolin, S. R., L. J. Runnels, C. A. Sawyer, K. J. Atcheson, and D. P. Gustafson. 1981. Resistance of porcine preimplantation embryos to pseudorabies virus. Am. J. Vet. Res. 42:1711–1712.

Bolin, S. R., L. J. Runnels, C. A. Sawyer, and D. P. Gustafson. 1982. Experimental transmission of pseudorabies virus in swine by embryo transfer. Am. J. Vet. Res. 43:278–280.

Bondioli, K. R., K. R. Gray, and J. B. Gibson. 1988. The effect of trypsin washing on post thaw viability of bovine embryos. Pp. 85–89 in Proceedings of the Seventh Annual Convention Meeting of the American Embryo Transfer Association. Reno, Nev.: American Embryo Transfer Association.

Bouillant, A. M. P., G. M. Ruckerbauer, M. D. Eaglesome, B. S. Samagh, E. L. Singh, W. C. D. Hare, and G. C. B. Randall. 1981. Attempts to isolate bovine leukemia and bovine syncytial viruses from blood, uterine flush fluid, unfertilized ova and embryos from infected donor cattle. Ann. Rech. Vet. 12(4):385–395.

Bowen, R. A., P. Spears, J. Storz, and G. E. Seidel, Jr. 1978a. Mechanisms of infertility in genital tract infections due to *Chlamydia psittaci* transmitted through contaminated semen. J. Infect. Dis. 138:95–98.

Bowen, R. A., J. Storz, and J. Leary. 1978b. Interaction of viral pathogens with preimplantation embryos. Theriogenology 9:88.

Bowen, R. A., T. H. Howard, and B. W. Pickett. 1982. Interaction of bluetongue virus with preimplantation embryos from mice and cattle. Am. J. Vet. Res. 43:1907–1911.

Bowen, R. A., T. H. Howard, R. P. Elsden, and G. E. Seidel, Jr. 1983. Embryo transfer from cattle infected with bluetongue virus. Am. J. Vet. Res. 44:1625–1628.

Bowen, R. A., R. P. Elsden, and G. E. Seidel, Jr. 1985. Infection of early bovine embryos with bovine herpesvirus-l. Am. J. Vet. Res. 46:1095–1097.

Chemineau, P., R. Procureur, Y. Cognie, P. C. Lefevre, A. Locatelli, and D. Chupin. 1986. Production, freezing and transfer of bluetongue virus-free goat embryos. Theriogenology 26:279–290.

del Campo, M. R., R. Tamayo, and C. H. del Campo. 1987. Embryo transfer from *brucellosis*-positive donors: A field trial. Theriogenology 27:221.

Di Giacomo, R. F., E. Studer, J. E. Evermann, and J. Evered. 1986. Embryo transfer and transmission of bovine leukosis virus in a dairy herd. J. Am. Vet. Med. Assoc. 188:827–828.

Dulac, G. C., and E. L. Singh. 1987. Embryo transfer as a means of controlling the transmission of viral infections. XII. The in vitro exposure of zona pellucida-intact porcine embryos to hog cholera virus. Theriogenology 29:1335–1341.

Eaglesome, M. D., W. C. D. Hare, and E. L. Singh. 1980. Embryo transfer: A discussion on its potential for infectious disease control based on a review of studies on infection of gametes and early embryos by various agents. Can. Vet. J. 21:106–112.

Eaglesome, M. D., D. Mitchell, K. J. Betteridge, G. C. B. Randall, E. L. Singh, B. S. Samagh, and W. C. D. Hare. 1982. Transfer of embryos from bovine leukemia virus-infected (BLV-positive) cattle to BLV-negative recipients. Preliminary results. Vet. Rec. 111(6):122–123.

Echternkamp, S. E., and R. R. Maurer. 1988. Capability of bovine embryos to develop in vitro after trypsin treatment and cryopreservation. Theriogenology 29:241.

Evermann, J. F., M. A. Faris, S. M. Niemi, and R. W. Wright. 1981. Pestivirus persistence and pathogenesis: Comparative diagnostic aspects of border disease virus of sheep and bovine viral diarrhea virus. Pp. 407–426 in Proceedings of the 24th Annual Meeting of the American Association of Veterinary Laboratory Diagnosticians. Columbia, Mo.: American Association of Veterinary Laboratory Diagnosticians.

Foote, W. C., J. W. Call, T. D. Bunch, and J. R. Pitcher. 1986. Embryo transfer in the control of transmission of scrapie in sheep and goats. Pp. 413–416 in Proceedings of the 90th Annual Meeting of the U.S. Animal Health Association. Richmond, Va.: U.S. Animal Health Association.

Gilbert, R. O., R. I. Coubrough, and K. E. Weiss. 1987. The transmission of bluetongue virus by embryo transfer in sheep. Theriogenology 27(3):527–540.

Guerin, B., J. P. Builly, P. Humblot, M. Nibart, and M. Thibier. 1988. Effets de la contamination experimentale in vitro des embryons de souris et de brebis par *Campylobacter fetus*. Bull. Acad. Vet. de France 61:63–78.

Hare, W. C. D. 1985. Diseases Transmissible by Semen and Embryo Transfer Techniques. Technical Series No. 4. Paris, France: Office International des Épizooties.

Hare, W. C. D., D. Mitchell, E. L. Singh, A. M. P. Bouillant, M. D. Eaglesome, G. M. Ruckerbauer, A. Bielanski, and G. C. B. Randall. 1985. Embryo transfer in relation to bovine leukemia virus (BLV) control and eradication. Can. Vet. J. 26:231–234.

Hare, W. C. D., A. J. Luedke, F. C. Thomas, R. A. Bowen, E. L. Singh, M. D. Eaglesome, G. C. B. Randall, and A. Bielanski. 1988. Nontransmission of bluetongue virus by embryos from bluetongue virus-infected sheep. Am. J. Vet. Res. 49:468–472.

Hasler, J. F., and A. M. Reinders. 1988. Pregnancy rate following transfer of bovine embryos treated with trypsin prior to freezing. Pp. 91–93 in Proceedings of the 7th Annual Convention Meeting of the American Embryo Transfer Association. Hastings, Neb.: American Embryo Transfer Association.

James, J. E. D., D. M. James, P. A. Martin, D. E. Reed, and D. L. Davis. 1983. Embryo transfer for conserving valuable genetic material from swine herds with pseudorabies. J. Am. Vet. Med. Assoc. 183:525–528.

Kaja, R. W., C. Olson, R. F. Rowe, R. H. Stauffacher, L. L. Strozinski, A. R. Hardie, and I. Bause. 1984. Establishment of a bovine leukosis virus-free dairy herd. J. Am. Vet. Med. Assoc. 184:184–185.

Lauerman, L. H., D. A. Stringfellow, P. H. Sparling, and L. M. Kaub. 1986. In vitro exposure of preimplantation bovine embryos to vesicular stomatitis virus. J. Clin. Microbiol. 24:380–383.

Mallek, Z., B. Guerin, M. Nibart, M. Parez, and M. Thibier. 1984. Consequences de la contamination in vitro des embryons de souris et de vaches par *Brucella abortus*. Bull. Acad. Vet. Fr. 57:479–490.

Mapletoft, R. J. 1987. The technology of embryo transfer. Pp. 2–40 in Proceedings of a Symposium on International Embryo Movement, W. C. D. Hare and S. M. Seidel, eds. Ottawa: Lowe-Martin.

McVicar, J. W., E. L. Singh, C. A. Mebus, and W. C. D. Hare. 1986. Embryo transfer as a means of controlling the transmission of viral infections. VIII. Failure to detect foot-and-mouth disease viral infectivity associated with embryos collected from infected donor cattle. Theriogenology 26:595–601.

Mebus, C. A., and E. L. Singh. 1988. Failure to transmit foot-and-mouth disease virus via bovine embryo transfer. Pp. 183–185 in Proceedings of the 92d Annual Meeting of the U.S. Animal Health Association. Richmond, Va.: U.S. Animal Health Association.

Olson, C., R. F. Rowe, and R. W. Kaja. 1982. Embryo transplantation and bovine leukosis virus. Preliminary report. Pp. 361–370 in Fourth International Symposium on Bovine Leukosis, O. C. Straub, ed. Boston: Martinus Nijhoff.

Parodi, A., G. Manet, A. Vaillaume, F. Crespau, B. Toma, and D. Levy. 1983. Transplantation embryonnaire et transmission de l'agent de la leucose bovine enzootique. Bull. Acad. Vet. Fr. 56:183–189.

Potter, M. L., R. E. Corstvet, C. R. Looney, R. W. Fulton, L. F. Archbald, and R. A. Godke. 1984. Evaluation of BVDV uptake by preimplantation embryos. Am. J. Vet. Res. 45:1778–1780.

Shelton, J. N. 1983. Prospects for the use of embryos in the control of disease and the transport of genotypes. Aust. Vet. J. 64:6–10.

Singh, E. L. 1987. The disease control potential of embryos. Theriogenology 27:9–20.

Singh, E. L., and F. C. Thomas. 1987a. Embryo transfer as a means of controlling the transmission of viral infections. IX. The in vitro exposure of zona pellucida-intact porcine embryos to swine vesicular disease virus. Theriogenology 27:443–449.

Singh, E. L., and F. C. Thomas. 1987b. Embryo transfer as a means of controlling the transmission of viral infections. XI. The in vitro exposure of bovine and porcine embryos to vesicular stomatitis virus. Theriogenology 28:691–697.

Singh, E. L., M. D. Eaglesome, F. C. Thomas, G. Papp-Vid, and W. C. D. Hare. 1982a. Embryo transfer as a means of controlling the transmission of viral infections. I. The in vitro exposure of preimplantation bovine embryos to Akabane, bluetongue and bovine viral diarrhea viruses. Theriogenology 17:437–444.

Singh, E. L., F. C. Thomas, M. D. Eaglesome, G. Papp-Vid, and W. C. D. Hare. 1982b. Embryo transfer as a means of controlling the transmission of viral infections. II. The in vitro exposure of preimplantation bovine embryos to infectious bovine rhinotracheitis virus. Theriogenology 18:133–140.

Singh, E. L., W. C. D. Hare, F. C. Thomas, and A. Bielanski. 1983. Embryo transfer as a means of controlling the transmission of viral infections. IV. Non-transmission of infectious bovine rhinotracheitis/infectious pustular vulvovaginitis virus from donors shedding virus. Theriogenology 20:169–176.

Singh, E. L., G. C. Dulac, and W. C. D. Hare. 1984. Embryo transfer as a means of controlling the transmission of viral infections. V. The in vitro exposure of zona pellucida-intact porcine embryos to African swine fever virus. Theriogenology 22:693–700.

Singh, E. L., J. W. McVicar, W. C. D. Hare, and C. A. Mebus. 1986. Embryo transfer as a means of controlling the transmission of viral infections. VII. The in vitro exposure of bovine and porcine embryos to foot-and-mouth disease virus. Theriogenology 26:587–593.

Singh, E. L., F. C. Thomas, W. C. D. Hare, and M. D. Eaglesome. 1987. Embryo transfer as a means of controlling the transmission of viral infections. X. The in vivo exposure of zona pellucida-intact porcine embryos to swine vesicular disease virus. Theriogenology 27:451–457.

Stringfellow, D. A., V. L. Howell, and P. R. Schurrenberger. 1982. Investigations into the potential for embryo transfer from *Brucella abortus* infected cows without transmission of infection. Theriogenology 18:733–743.

Stringfellow, D. A., C. M. Scanlan, S. J. Hannon, V. S. Panagala, B. W. Gray, and P. A. Galik. 1983. Culture of uterine flushings, cervical mucus, and udder secretions collected post-abortion from heifers artificially exposed to *Brucella abortus*. Theriogenology 20:77–83.

Stringfellow, D. A., C. M. Scanlan, R. R. Brown, G. B. Meadows, B. W. Gray, and R. R. Young-White. 1984. Culture of bovine embryos after in vitro exposure to *Brucella abortus*. Theriogenology 21:1005–1012.

Stringfellow, D. A., V. S. Panangala, and P. A. Galik. 1988. Recovery and culture of ova from *Brucella abortus* infected cows. Theriogenology 29:1105–1112.

Thibier, M., and M. Nibart. 1987. Disease control and embryo importations. Theriogenology 27:37–47.

Thomas, F. C., E. L. Singh, and W. C. D. Hare. 1983. Embryo transfer as a means of controlling viral infections. III. Non-transmission of bluetongue virus from viremic cattle. Theriogenology 19:425–431.

Thomas, F. C., E. L. Singh, and W. C. D. Hare. 1985. Embryo transfer as a means of controlling viral infections. VI. Bluetongue virus-free calves from infectious semen. Theriogenology 24:345–350.

Thomson, M. S., D. A. Stringfellow, and L. H. Lauerman. 1987. In vitro exposure of reimplantation bovine embryos to *Haemophilus somnus*. Theriogenology 27:287.

Voelkel, S. A., K. W. Stuckey, C. R. Looney, F. M. Enright, P. E. Humes, and R. A. Godke. 1983. An attempt to isolate *Brucella abortus* from uterine flushings of superovulated donor cattle. Theriogenology 19:355–366.

Wolfe, D. F., K. E. Nusbaum, L. H. Lauerman, P. W. Mysinger, M. G. Riddell, L. S. Putman, and T. A. Powe. 1987. Embryo transfer from goats seropositive for caprine arthritis-encephalitis virus. Theriogenology 28:307–316.

Wrathall, A. E., and W. L. Mengeling. 1979a. Effect of porcine parvovirus on development of fertilized pig eggs in vitro. Br. Vet. J. 135:249–254.

Wrathall, A. E., and W. L. Mengeling. 1979b. Effect of transferring parvovirus-infected fertilized pig eggs into seronegative gilts. Br. Vet. J. 135:255–261.

ANNEX B-1
Sanitary Techniques for Handling Embryos

This annex describes the manner in which embryos should be treated between collection and transfer in order to preclude the possibility of transferring disease agents along with the embryo. Although the technique of embryo transfer can be used to limit the disease transmission potential of embryos, if the technique is improperly carried out, it could be used to introduce disease. The procedures described are recommended by the International Embryo Transfer Society and endorsed by the Office International des Épizooties. They are based on research that has been carried out on the transmissibility of disease pathogens by the embryo. This research has shown that the presence of an intact zona pellucida is an important structure in maintaining the disease-free status of the embryo itself and that proper washing can render embryos free of many pathogens. Although the presence of the zona pellucida is an important barrier that prevents many pathogens from gaining access to the embryo, under certain circumstances this structure can have a negative impact on the health status of the embryo. That is because some disease agents have been found to adhere to the zona pellucida so firmly that proper washing will not remove them. Thus, although the embryo itself would still be disease free, transfer of an embryo with a pathogen on the zona pellucida could lead to infection of the recipient and, ultimately, infection of the embryo or fetus. Research has shown, however, that many of the disease agents that adhere to the zona pellucida can be removed by briefly treating the embryos with trypsin. This treatment, which minimizes the risk of disease transmission, does not affect embryo viability adversely.

EMBRYO COLLECTION

Collections must be carried out using strict aseptic procedures. The vulva of the donor should be scrubbed with an antiseptic and then rinsed well. All collection equipment (catheters, recovery tubing, petri dishes, and so on) should be sterile. In addition, all media, solutions, and sera that come into contact with the embryo must be free of contaminants and living microorganisms.

EMBRYO WASHING

The recommended washing procedure involves transferring embryos, in groups of 10 or fewer, through 10 changes of sterile medium containing antibiotics. A fresh, sterile micropipette must be used to transfer the embryos to each of the washes; each wash must constitute a hundredfold dilution of the previous wash. The embryos must be gently agitated in each of the washes; as soon as that has been carried out, they can be moved to the next wash. Only embryos from the same donor are washed together.

Trypsin Treatment of Embryos

If it is deemed necessary that embryos be treated with trypsin, the following combined washing and trypsin treatment should be followed. Embryos are transferred through five washes of sterile phosphate-buffered saline without calcium (Ca^{++}) and magnesium (Mg^{++}) but with antibiotics and 0.4 percent bovine serum albumin. It is important to remove serum and divalent cations before the embryos are exposed to trypsin to insure the biological activity of this enzyme. The embryos are then exposed to two washes of trypsin, pH 7.6 to 7.8, for a total time in the trypsin of 60 to 90 seconds. Sterile trypsin (trypsin 1:250 that has an activity such that 1 g will hydrolyze 250 g of casein, at 25°C and pH of 7.6, in 10 minutes) in Hanks' balanced salt solution, without Ca^{++} and Mg^{++}, is used at a concentration of 0.25 percent. After the trypsin treatment the embryos are transferred through five washes of phosphate-buffered saline containing Ca^{++}, Mg^{++}, antibiotics, and 2 percent serum. Serum is a potent inhibitor of trypsin and its inclusion in the last five washes ensures that the action of trypsin will be stopped.

Inspection of Embryos for Intactness of Zona Pellucida

All embryos must have a zona pellucida that is intact and free of adherent material. To ensure this, each embryo must be examined over its entire surface area at not less than 50 magnification. Embryos are gently rolled in the dish so that all surfaces on the zona pellucida can be examined. This evaluation should take place after washing and before embryo micromanipulation or cryopreservation.

Micromanipulation of Embryos

As long as embryos are micromanipulated after being washed properly, the actual micromanipulation itself should not alter the health status of the embryo. This might not hold true, however, if foreign zona pellucidae were used in the embryo micromanipulation process. Under these circumstances it would be essential that the foreign zona pellucidae had been obtained from properly washed embryos/eggs and that they were not derived from a donor infected (seropositive) with any pathogen that is known to adhere to the zona pellucida.

CRYOPRESERVATION AND THAWING OF EMBRYOS

All solutions, cryoprotectants, and containers must be sterile, and a sterile technique must be used throughout these procedures.

Embryo Identification

Proper identification is essential if embryos are to be related to their health certificates. Thus, containers (straws or ampoules) of frozen embryos must be labeled with an initial code that identifies the embryo transfer company code, the breed code, and the donor cow and sire registration numbers. In addition, the straw or ampoule must also be marked with the freezing date (year/month/day), a straw or ampoule number (for a given collection), and the number of embryos it contains.

Goblets and canes holding straws or ampoules of frozen embryos must be labeled with the cane number, the embryo transfer company code, the date of freezing, the breed code, and the registered name and number of the donor female and male.

EMBRYO TRANSFER

The transfer of embryos is carried out using strict aseptic procedures.

CONCLUSION

In terms of disease control, the safest form of genetic material is embryos. If embryos are properly processed, the evidence indicates that disease transmission will rarely, if ever, occur.

What may turn out to be more important to the health of the embryo (fetus) than the disease status of the genetic dam and sire is the health of the recipient. It is essential that embryos be transferred to recipients that are free from infectious diseases that can be transmitted "in utero." Otherwise, the enormous benefits of embryo transfer in terms of disease control will be lost.

ANNEX B-2
Tables Summarizing Research Data

The following seven tables summarize the research data presented in Annex B-1.

TABLE B2-1 Embryonic Development and Infectivity of Zona-Intact Embryos After Pathogen Exposure and Washing

Pathogen	Titer of Pathogen	Number of Embryos Exposed	Effect on Development	Percentage of Embryos Carrying Pathogen	Reference(s)
Bovine embryos					
AV	10^6 pfu/ml	80	None detected	0	Singh et al., 1982a
B. abortus	10^1 to 10^{14} cells/ml	96	Decreased viability	0	Mallek et al., 1984; Stringfellow et al., 1984
BLV	Cocultivation infected cells	12	None detected	0	Hare et al., 1985
BTV	10^2 to 10^7 pfu/ml	120	None detected	0	Bowen et al., 1982, 1985
BVDV	10^4 to 10^5 $TCID_{50}$/ml	122	None detected	0	Singh et al., 1982a; Potter et al., 1984
FMDV	10^6 pfu/ml	169	None detected	0	Singh et al., 1986
H. somnus	10^{10} organisms/ml	38[a]	Decreased viability	26	Thomson et al., 1987
H. somnus	10^{10} organisms/ml	20[b]	Not tested	0	Thomson et al., 1987
IBRV	10^6 to 10^8 $TCID_{50}$/ml	83	None detected	70	Singh et al., 1982b
RPV	10^5 $TCID_{50}$/ml	61	None detected	3	Dulac and Singh, unpublished data
VSV	10^5 to 10^7 pfu/ml	130	None detected	36	Lauerman et al., 1986; Singh et al., 1987

Porcine embryos					
ASFV	10^6 $HADD_{50}$/ml	80	None detected	95	Singh et al., 1984
FMDV	10^6 pfu/ml	194	None detected	5	Mebus et al., 1988; Singh et al., 1986
HCV	10^4 to 10^6 FFU/ml	171	None detected	97	Dulac and Singh, 1987
PP	10^4 $TCID_{50}$/ml	38	Decreased viability	100	Wrathall and Mengeling, 1979a
PrV	10^4 to 10^8 $TCID_{50}$/ml	282	None detected	24	Bolin et al., 1982; Singh and Thomas, unpublished data
SVDV	10^7 pfu/ml	176	None detected	100	Singh and Thomas, 1987a
VSV	10^5 to 10^6 pfu/ml	97	None detected	85	Singh and Thomas, 1987b
Ovine embryos					
BVDV	10^2 pfu/ml	49	None detected	0	Evermann et al., 1981
C. fetus	0.1 to 100 ID_{50}	164	Not tested	0	Guerin et al., 1988

NOTE: ASFV = African swine fever virus; AV = Akabane virus; *B. abortus* = *Brucella abortus;* BLV = bovine leukosis virus; BTV = bluetongue virus; BVDV = bovine viral diarrhea virus; *C. fetus* = *Campylobacter fetus;* FMDV = foot-and-mouth disease virus; HCV = hog cholera virus; *H. somnus* = *Haemophilus somnus;* IBRV = infectious bovine rhinotracheitis virus; PP = porcine parvovirus; PrV = pseudorabies virus; RPV = rinderpest virus; SVDV = swine vesicular disease virus; VSV = vesicular stomatitis virus; pfu = plaque-forming units; ml = milliliters; $TCID_{50}$ = tissue culture infective dose 50 percent; $HADD_{50}$ = hemadsorption dose 50 percent; FFU = focus-forming units; ID_{50} = infective dose 50 percent.

[a]Embryos not exposed to antibiotics.
[b]Embryos exposed to antibiotics.

TABLE B2-2 Effect of Trypsin Treatment on Pathogen-Exposed, Zona-Intact Embryos

Pathogen	Titer of Pathogen	Number of Embryos Treated with Trypsin	Percentage of Embryos Carrying Pathogen	Reference(s)
Bovine embryos				
IBRV	10^7 $TCID_{50}$/ml	32	0	Singh et al., 1982b
VSV	10^7 pfu/ml	38	0	Singh and Thomas, 1987b
Porcine embryos				
ASFV	10^6 $HADD_{50}$/ml	126	30	Singh et al., 1984
HCV	$<10^6$ FFU/ml	114	0	Dulac and Singh, 1987
HCV	$>10^6$ FFU/ml	118	24	Dulac and Singh, 1987
PrV	10^6 $TCID_{50}$/ml	45	0	Singh and Thomas, unpublished data
SVDV	10^7 pfu/ml	100	64	Singh and Thomas, 1987a
VSV	10^6 pfu/ml	61	0	Singh and Thomas, 1987b

NOTE: Embryos were treated with 0.25 percent trypsin, pH 7.6 to 7.8 for 60 to 90 seconds. When this was ineffective in removing infectivity (ASFV and SVDV), longer periods (2 to 15 minutes) of trypsin treatment were used with limited success. ASFV = African swine fever virus; HCV = hog cholera virus; IBRV = infectious bovine rhinotracheitis virus; PrV = pseudorabies virus; SVDV = swine vesicular disease virus; VSV = vesicular stomatitis virus; pfu = plaque-forming units; ml = milliliters; $TCID_{50}$ = tissue culture infective dose 50 percent; $HADD_{50}$ = hemadsorption dose 50 percent; FFU = focus-forming units.

TABLE B2-3 Embryonic Development and Infectivity of Zona-Free Embryos After Pathogen Exposure and Washing

Pathogen	Titer of Pathogen	Number of Embryos Exposed	Effect on Development	Assay of Embryos	Reference(s)
Bovine embryos					
B. abortus	10^1 to 10^5 cells/ml	3	Decreased viability	Negative	Mallek et al., 1984; Stringfellow et al., 1984
BLV	Cocultivation infected cells	15	None detected	Negative	Hare et al., 1985
BPV	—	—	None detected	Negative	Bowen et al., 1978b
BTV	10^5 pfu/ml	18	Embryo death	Positive	Bolin et al., 1982
BVDV	10^5 pfu/ml	26	None detected	Negative	Potter et al., 1984
FMDV	10^6 pfu/ml	42	None detected	Positive	Singh et al., 1986
H. somnus	10^{10} organisms/ml	9	Decreased viability	Negative	Thomson et al., 1987
IBRV	10^7 $TCID_{50}$/ml	53	Embryo death	Positive	Bowen et al., 1985
Porcine embryos					
PrV	10^4 to 10^8 $TCID_{50}$/ml	46	Not tested	Negative	Bolin et al., 1981
Ovine embryos					
BVDV	10^3 pfu/ml	23	None detected	Negative	Evermann et al., 1981

NOTES: A dash (—) indicates unavailable information. *B. abortus* = *Brucella abortus;* BLV = bovine leukosis virus; BPV = bovine parvovirus; BTV = bluetongue virus; BVDV = bovine viral diarrhea virus; FMDV = foot-and-mouth disease virus; *H. somnus* = *Haemophilus somnus;* IBRV = infectious bovine rhinotracheitis virus; PrV = pseudorabies virus; pfu = plaque-forming units; ml = milliliters; $TCID_{50}$ = tissue culture infective dose 50 percent.

TABLE B2-4 Transfer of Embryos from Seropositive or Infected Donors

Donors	Serological Number of Embryos Transferred	Number of Recipients	Status of Offspring and Recipients	Reference(s)
Bovine embryos				
B. abortus seropositive	39	39	Negative	del Campo et al., 1987
BLV seropositive	596	≤596	Negative	Eaglesome et al., 1982; Olson et al., 1982; Parodi et al., 1983; Kaja et al., 1984; Hare et al., 1985; DiGiacomo et al., 1986
BLV seropositive	1,500[a]	<1,500	Negative	Thibier and Nibart, 1987
BTV infected	334	330	Negative	Acree, 1988; Bowen et al., 1983; Thomas et al., 1983, 1985
FMDV infected	149	112	Negative	McVicar et al., 1986; Mebus and Singh, 1988
IBRV infected	64[b]	49	Negative	Singh et al., 1983
IBRV seropositive	>1,500[a]	<1,500	Negative	Thibier and Nibart, 1987
RPV infected	17	15	Negative	Singh and Mebus, unpublished data
Porcine embryos				
HCV infected	385	17	Negative	Singh and Mebus, unpublished data
FMDV infected	436	13	Negative	Singh and Mebus, unpublished data

PrV infected intranasally and intrauterinely	70	5	Two of five were seroconverted	Bolin et al., 1982
PrV infected intranasally	45	3	Negative	Bolin et al., 1982
PrV seropositive	805	34	Negative	James et al., 1983
SVDV infected	205	9	Negative	Singh et al., 1987
Ovine embryos				
BTV infected	49	27	Negative	Hare et al., 1988
BTV infected	20	15	Two of five were seroconverted	Gilbert et al., 1987
Scrapie infected	?[c]		Negative	Foote et al., 1986
Caprine embryos				
BTV seropositive	63[d]	19	Negative	Chemineau et al., 1986
CAEV seropositive	16	8	Negative	Wolfe et al., 1987

NOTE: *B. abortus* = *Brucella abortus;* BLV = bovine leukosis virus; BTV = bluetongue virus; CAEV = Caprine arthritis and encephalitis virus; FMDV = foot-and-mouth disease virus; HCV = hog cholera virus; IBRV = infectious bovine rhinotracheitis virus; PrV = pseudorabies virus; RPV = rinderpest virus; SVDV = swine vesicular disease virus.

[a]These embryos were derived from herds in which the majority were BLV and IBRV seropositive.

[b]Embryos were treated with trypsin prior to transfer.

[c]The transfers have produced 69 lambs, which have remained scrapie uninfected (approximately 5 years old). Scrapie was diagnosed in the positive controls around 40 months.

[d]The 63 embryos had been deep frozen prior to transfer and were derived from a herd in which 47 percent were serologically positive for BTV.

TABLE B2-5 Assay of Embryos Derived from Seropositive or Infected Donors

Status of Donor	Number of Embryos Collected	Number of Eggs Collected	Assay of Embryos	Reference(s)
Bovine embryos				
B. abortus infected	309	82	Negative	Voelkel et al., 1983; Barrios et al., 1988; Stringfellow et al., 1988
BLV seropositive	60	26	Negative	Bouillant et al., 1981
BTV infected	51	14	Negative	Bowen et al., 1983
BVDV seropositive	2	21	Negative	Singh et al., 1982a
C. psittaci infected	5		Negative	Bowen et al., 1978a
FMDV infected	372	64	Negative	McVicar et al., 1986; Mebus and Singh, 1988
IBRV infected	63[a]		Negative	Singh et al., 1983
IBRV seropositive	13	18	Negative	Singh et al., 1982b
RPV infected	170		Negative	Singh and Mebus, unpublished data
Bred with infected semen	47		Positive	Singh, unpublished data

Porcine embryos				
ASFV infected	24		Negative	Singh and Dulac, unpublished data
HCV infected	19	74	Negative	Singh and Mebus, unpublished data
FMDV infected	177	104	Negative	Singh and Dulac, unpublished data
SVDV infected	17	95	Negative	Singh et al., 1987

NOTE: ASFV = African swine fever virus; *B. abortus* = *Brucella abortus;* BLV = bovine leukosis virus; BTV = bluetongue virus; BVDV = bovine viral diarrhea virus; *C. psittaci* = *Chlamydia psittaci;* FMDV = foot-and-mouth disease virus; HCV = hog cholera virus; IBRV = infectious bovine rhinotracheitis virus; RPV = rinderpest virus; SVDV = swine vesicular disease virus.

[a]These embryos were trypsin treated prior to assay.

TABLE B2-6 Assay of Uterine Flush Fluids from Seropositive or Infected Donors

Status of Donor Flushed	Number of Positive Flushes	Reference(s)
Bovine embryos		
BLV seropositive	4/25	Bouillant et al., 1981
BTV infected	12/30	Bowen et al., 1983; Thomas et al., 1983
BTV-infected semen (bred with)	0/4	Thomas et al., 1985
FMDV infected	15/22	McVicar et al., 1986; Mebus and Singh, 1988
IBRV infected	9/33	Singh et al., 1982b
B. abortus infected	9/116	Stringfellow et al., 1982, 1983, 1988; Voelkel et al., 1983
Porcine embryos		
FMDV infected	4/15	Mebus and Singh, 1988
HCV infected	6/33	Singh and Dulac, unpublished data
SVDV infected	8/17	Singh et al., 1987
Caprine embryos		
CAEV seropositive	0/12	Wolfe et al., 1987

NOTE: *B. abortus* = *Brucella abortus;* BLV = bovine leukosis virus; BTV = bluetongue virus; CAEV = caprine arthritis and encephalitis virus; FMDV = foot-and-mouth disease virus; HCV = hog cholera virus; IBRV = infectious bovine rhinotracheitis virus; SVDV = swine vesicular disease virus.

TABLE B2-7 Transfer of Embryos Exposed to Pathogens In Vitro

Pathogen	Titer of Exposure Pathogen	Number of Embryos Exposed and Transferred	Number of Recipients	Serological Status of Offspring and Recipients	Reference(s)
Bovine embryos					
BLV	Cocultivation infected cells	48	3	Negative	Hare et al., 1985
IBRV	10^7 $TCID_{50}$/ml	38[a]	22	Negative	Singh, unpublished data
IBRV	10^7 $TCID_{50}$/ml	8	4	One of four positive	Singh, unpublished data
Porcine embryos					
PP	10^4 $TCID_{50}$/ml	46	4	Four of four positive	Wrathall and Mengeling, 1979b
PrV	10^4 $TCID_{50}$/ml	45	4	Negative	Bolin et al., 1982
PrV	10^8 $TCID_{50}$/ml	79	5	Five of five positive	Bolin et al., 1982
PrV	10^7 $TCID_{50}$/ml	223[a]	10	Negative	Singh and Thomas, unpublished data
PrV	10^7 $TCID_{50}$/ml	117	6	Six of six positive	Singh and Thomas, unpublished data
Ovine embryos					
BTV	10^4 pfu/ml	13	3	Positive	Gilbert et al., 1987

NOTE: BLV = bovine leukosis virus; BTV = bluetongue virus; IBRV = infectious bovine rhinotracheitis virus; PP = porcine parvovirus; PrV = pseudorabies virus; pfu = plaque-forming units; ml = milliliters; $TCID_{50}$ = tissue culture infective dose 50 percent.

[a]These embryos were trypsin treated prior to transfer.

APPENDIX

C

Animal Genetic Resources: Sperm

Ian Parsonson

The importance of using semen in the dissemination of the genetic diversity of domesticated animal species is well recognized. Increases in highly desirable genetic characteristics and in the productivity of the various animal species have established artificial breeding (AB) as an efficient and reliable method in animal husbandry.

Technical progress in the cryopreservation of semen and in modern handling, storing, and transport systems have made international trade in animal semen an important segment of the commerce of many countries. These technical achievements have been a critical factor in the improvement of livestock worldwide. The portability and keeping qualities of cryopreserved semen have provided access to gene pools around the world. As techniques are developed for the collection, handling, and storage of semen from additional animal species not widely used at present, further expansion of world genetic resources will occur (Seidel, 1986).

Authorities responsible for disease control within a country usually require specific health standards for the donors standing at AB centers, as well as general standards for the operation of the centers and for the handling, storage, and quality of the gametes produced from the centers. International trade requires standards and tests for

---

Ian Parsonson was assistant chief of the Australian Animal Health Laboratory at the Commonwealth Scientific and Industrial Research Organization. He is a member of the Australian Commonwealth Government Genetic Manipulation Advisory Committee.

semen, which are set out in protocols by the importing countries. These protocols and regulations define the standards for the donor animals, the AB centers, and the bacteriological control of the product. Generally, the quality control requirements set high standards for most AB establishments. The standards include disease control conditions for entry of animals into centers, for health standards for animals once in the centers, and for sale of semen from donors at the AB center.

Standards for the importation of semen from countries in Europe, North America, and Australasia are comparable; they include tests for disease agents known to be exotic to the importing country. The advantage of cryopreserved semen is that it can be held pending verification of the accredited health status of the donors before being released for sale. An additional advantage of holding cryopreserved semen is that progeny testing can be conducted to evaluate genetic gain or to detect genetic defects before general release of the semen samples.

Transmission of disease agents in semen or in spermatozoa has been seen as potentially the most likely method of spread of infectious disease both domestically and internationally. The realization that cryopreservation, antibiotics, and semen extenders such as milk and egg yolk citrate also assisted in the preservation of many pathogens led to the development of health standards for donors and codes of practice for AB centers. Adherence to protocols for the health status of animals in AB centers and use of strict aseptic collection, handling, and preservation methods can lead to a high-quality, minimally infected product. Artificial breeding also provides a method of disease control through the use of specific pathogen free semen.

Artificial breeding centers are the basis of national disease control programs in many countries and are usually a requirement for an export trade in semen. The Office International des Épizooties (OIE) has developed recommendations for certification of AB centers on the basis of sanitary codes for control of disease and standards for processing semen (Office International des Épizooties, 1986). Three disease-free categories—regional freedom, herd freedom, and individual donor and its gametes freedom from specific diseases—form the basis for international trade in animal gametes and in particular semen.

MAJOR SPECIES OF DOMESTIC LIVESTOCK INVOLVED IN INTERNATIONAL GERMPLASM TRADE

The major species of domestic livestock involved in international trade in germplasm are cattle, sheep, pigs, goats, horses, and poultry.

Species involved at a much reduced level include buffalo, deer, dogs, and various types of zoological animals. In addition, there is some trade in insect germplasm (bees), in fish germplasm, and in certain lines of laboratory animals, usually as frozen embryos. The major species used in AB programs throughout, however, is cattle (*Bos taurus* and *Bos indicus*), and it is through the production of semen from this species that the advantages and problems of artificial breeding have been identified.

When bovine semen was first used for artificial breeding, the distribution of the semen was restricted to local regions because of the limited viability of the product and the lack of mobility of the inseminators. There was a relationship between the fertility of particular bulls and the pregnancy rate of cows inseminated with their semen, and soon, bulls whose nonreturn rates were lower than expected standards were detected. It also soon became obvious that certain venereal diseases were being transmitted with the semen. Some of the diseases, such as brucellosis (*Brucella abortus*) and trichomoniasis (*Trichomonas foetus*), had been identified as causes of reproductive failure, and attempts were made to diagnose infected bulls. As bovine semen became an article of commerce, quality control standards became mandatory in many countries. Several other infectious agents affecting fertility were added to the growing list of pathogens that might be present in semen, including Campylobacteriosis (*Campylobacter fetus* var. *venerealis* and *C. intermedius*).

The cryopreservation of extended semen added to the problem in that correctly frozen semen also preserved the microorganisms that were present as contaminants. Antibiotics were added to semen extenders not so much to eradicate the microorganisms present, but rather to reduce their potential to cause harm. Along with these steps, research into improved aseptic collection and handling of fresh semen showed the importance of careful attention to temperature and temperature gradients for cryopreservation.

ARTIFICIAL BREEDING CENTERS AND DONOR ANIMAL REQUIREMENTS

The Australian Standing Committee on Agriculture has required that AB centers comply with Australian standards to be eligible for registration with the Australian Quarantine Inspection Service for the export of semen. Bovine semen production centers are licensed under state legislation to produce semen for sale from bulls kept within the center and judged to be disease free according to minimum standards set out in the *Minimum Health Standards for Stock Standing at*

Licensed or Approved Artificial Breeding Centers in Australia (Australian Quarantine and Inspection Service, 1988).

A center is a quarantine area and is under veterinary supervision. The center must have facilities within the fully health-tested area for accommodation of stock and for collection of semen from the stock. It must have a processing laboratory and storage facilities for semen derived from the stock. Other requirements include providing for treatment of sick animals, introducing new stock without endangering those already present, and maintaining the center as a strict quarantine area.

Prior to entering the center, cattle are examined for freedom from infectious disease and, as possible, tested for hereditary defects (for example, mannosidase tests for Angus cattle and crosses of Angus extraction) to detect carriers. The health tests require freedom from brucellosis (*Brucella abortus*), tuberculosis (*Mycobacterium bovis*), and leptospirosis (*L. pomona* and *L. hardjo*). Treatment with dihydrostreptomycin is also required but only after tests have been made for freedom from campylobacteriosis and trichomoniasis. Cattle cannot be derived from a property on which Johnes disease (*Mycobacterium paratuberculosis*) has been diagnosed within the previous 5 years. In addition, they must have negative tests for paratuberculosis. Cattle must also be free of antibodies to enzootic bovine leukosis, and the property must have a history of freedom from the disease. All of these diseases are listed as "B" or "Others" on the OIE disease lists (discussed below). Once the animals are in the fully health-tested area of the licensed center, they must undergo annual testing for freedom from the above-mentioned diseases.

Semen production centers in Australia must comply with the Australian and OIE standards to be eligible to export semen. By definition, a licensed bovine semen production center is licensed under state legislation to produce semen for sale from bulls kept within the center and judged to be disease free according to the minimum standards described by the Australian Quarantine Inspection Service (1988).

Similar requirements are laid out in regulations governing AB centers in the United States, Canada, Great Britain, and New Zealand, and they form part of the OIE standards for artificial breeding. Additional disease control tests are required of countries in which the diseases on the OIE's List A (discussed below) are present. Even for many of these important disease agents, however, information on their transmission in semen is minimal. Often, the evidence is obtained during the viremic phase of the disease, when the agent is at its highest titer in the blood. Evidence of the persistent excretion of viruses in semen is available only for a few diseases, and it may be an insignificant problem, particularly when health standards and tests are applied to donors prior to and following semen collection.

Of the diseases listed by the OIE (Hare, 1985), lists A, B, and Others include a number of agents that have been isolated in semen from bulls, rams, bucks, and boars. The evidence, however, is often minimal or restricted in scope, and in some cases confusing reports on individual agents may be given. For many infectious agents there is an obvious need for further studies to identify their potential to transmit disease through the genital route.

Current techniques for artificial breeding provide access to the cervix and uterus, which is not presented normally in natural mating. It is the necessity to deliver semen through this route in many species, or through laparoscopy in some species, that enables the agents to bypass the immunological and physiological defenses of the female.

CONTROL OF COLLECTION, HANDLING, PRESERVATION, AND STORAGE OF SEMEN

Standards have been developed in many countries for donor animals and the collection and storage of semen. With the advent of liquid nitrogen cryopreservation and the development of extender recipes and antibiotic regimes, the quality of preserved semen has improved. Even with all previous research and continuing improvement in techniques for collection, handling, and storage, semen remains a product contaminated with microorganisms. Further attempts to reduce the level of contamination have concentrated on the health of the donor animals, their freedom from specific diseases, and as far as possible, asepsis during collection and subsequent handling of semen. The latter includes ensuring the sterility of extender liquids and the addition of a suitable antibiotic cover. Because of the need to remove or control mycoplasmas, the antibiotic cover has now been changed in many countries and will soon become a standard component of semen for international use. In the United States, the use of penicillin, dihydrostreptomycin, and polymyxin B sulphate has been replaced in the minimum requirements of the Certified Semen Services (CSS) with the new combination of tylosin, gentamicin, lincomycin, and spectinomycin (Doak, 1986). The change to the use of this antibiotic combination has shown that processing procedures, extender composition, and antibiotic combinations may affect the efficacy of microbial control or fertility.

The CSS has listed the following requirements (Doak, 1986):

- Extender must be of a type approved by the CSS.
- Contact between antibiotics and raw semen must be for less than 3 minutes.
- Cooling of semen and nonglycerol fraction of less than 2 hours to 5°C.
- Glycerol cannot be added as an extender component until after cooling to 5°C.

The new antibiotics and procedures have been accepted in the United States for intra- and interstate movement of semen for artificial breeding. International movement of semen will require exporting countries to comply with the changes. Many countries outside the United States are moving to adopt similar changes (Shin, 1986), and such changes may be adopted as part of the quarantine protocols for international movement of germplasm.

MAJOR DISEASES OF CONCERN IN ARTIFICIAL BREEDING

List A Agents

The major diseases of concern in the international transfer of gametes for artificial breeding are mostly those caused by viruses. The IOE's List A Agents are communicable diseases that (1) have the potential for very serious and rapid spread irrespective of borders, (2) are of serious socioeconomic or public health consequences, and (3) are of major importance in the international trade of livestock and livestock products (Gibbs, 1981; Hare, 1985; Odend'hal, 1983; Parez, 1985). Some of these diseases are described below.

Foot-and-Mouth Disease Virus

The most important virus disease on the OIE's List A is that caused by the foot-and-mouth disease virus (FMDV) because of its presence on all continents, except Australia and North America, and because of the number of animal species affected by FMDV. Cottral et al. (1968) identified the presence of FMDV in semen of bulls and its subsequent transmission by artificial insemination. This finding was confirmed by Sellers et al. (1968), who also showed that bulls exposed by indirect-contact infection had high titers of FMDV in the semen before the lesions of clinical disease appeared. These findings precluded the use of semen from FMDV-infected countries for a number of years. As new testing systems were developed, protocols were also developed by some countries to allow international movement of semen. Acceptance of these protocols has enabled considerable headway to be made in relation to other disease agents on List A, such as bluetongue virus (BTV).

Bluetongue Virus

BTV can infect a wide range of ruminant species. During the past 10 years, research into the excretion of BTV in bovine and ram semen has shown that, in the vast majority of cases, BTV is only shed during the period of viremia, and then at a titer that is considerably lower than in the blood (Bowen et al., 1983; Breckon et al., 1980; Parsonson et al., 1981b, 1987a,b). When semen contaminated with BTV is instilled into the cervix and uterus of susceptible cattle, it usually causes infection, generally without clinical signs, and it appears to have little or no effect on subsequent pregnancy (Bowen and Howard, 1984; Parsonson et al., 1987a,b). However, when cattle were experimentally infected with insect-passaged (*Culicoides variipennis*) BTV serotypes 11 and 17, congenitally deformed calves were produced (Luedke et al., 1977). Similar experiments have not produced infected calves, nor was serological or virological evidence of transplacental transfer of BTV found in those studies (Parsonson, 1990; Parsonson et al., 1987b). In a study in which bovine fetuses were inoculated in utero with BTV at 125 days gestation, MacLachlan et al. (1984) showed that the virus did not persist in the fetus at birth. Thus, fetuses infected in utero are unlikely to be reservoirs for the virus.

Additional studies on the fetal transmission of BTV seem unnecessary in light of the mounting evidence against transmission by this route. There is need, however, for further studies of the initial reproductive stages in sheep and cattle to ensure that BTV is not a cause of reproductive losses in endemic areas. Despite the experimental evidence, Osburn (1986) suggested a need for further studies in dairy cattle in California.

Rinderpest Virus

Although rinderpest virus can infect a wide range of ruminant species, there is very little evidence for its spread through artificial breeding. Scott (1964), however, reported isolation of the virus from all secretions and excretions of infected animals, and Plowright (1968) cited Curasson (1921) and Jacotot (1931) as finding the virus in vaginal secretions for 5 to 12 weeks after abortion had occurred. Because rinderpest and peste des petits in ruminants cause obvious disease on a herd basis, it is unlikely that semen or ova for artificial breeding would be collected from animals in an area where an outbreak has occurred.

In an extensive series of tables, Scott (1964) has detailed animal species known to be infected with rinderpest virus. The lists are important when considering species that are endangered or at risk.

Lumpy Skin Disease (Neethling Virus)

In experimentally infected animals, Weiss (1968) demonstrated the presence of lumpy skin disease virus (LSDV) in bull semen for 22 days following fever and generalized skin lesions. Weiss, however, also demonstrated the presence of virus in semen at days 7 and 12 in bulls undergoing inapparent infections. Almost certainly, virus would be present in secretions of cows from which embryos were collected during infection. The similarity between LSDV and capripox has been noted in several reports, and LSDV has been classified in the genus *Capripoxvirus* by Fenner (1976).

Sheep and Goat Pox Virus

These viruses were originally thought to be species specific. However, Kitching and Taylor (1985), however, have shown a close serological antigenic relationship, and cross-protection studies have supported that finding. As a result, Kitching and Taylor have suggested that the term *capripox* be used in the future, irrespective of the species involved.

Because of the similarities between these viruses and LSDV, excretion in both semen and vaginal discharges are extremely likely during infection of donors. Because collection of semen or ova would be unlikely during the occurrence of clinical disease, there should be little chance of venereal transmission. This group of viruses is extremely resistant to environmental factors, however, and all precautions to avoid venereal transmission must be taken. A suitable attenuated virus vaccine for capripox, the 0240LT/1 BHK/2 LT/4 strain of Kenyan sheep and goat pox virus, has been shown to be stable and safe and to provide substantial protection for at least 1 year (Kitching et al., 1987). Semen and embryos could therefore be collected from donor animals vaccinated against *Capripoxvirus*.

Rift Valley Fever Virus

Rift Valley fever virus (RVFV) is an arthropod-borne virus that can also be spread by contamination of the environment and by aerosols. It is a particularly difficult agent to control. Vaccination is carried out in some areas of the African continent where epizootics have occurred and in endemic areas where sporadic outbreaks cause problems to livestock. Although RVFV causes abortions in pregnant animals of many species, including humans, and although it can be fatal to neonatal and young animals, there is a paucity of information on its effect on reproduction. This lack of information may be related to the severity of the clinical condition and the insignificant role that reproductive disorders would play in an outbreak of the disease. However, there is a need to know how to avoid transfer of this agent in gametes originating in endemic areas of Africa and destined for international trade. Several wild animal species from Africa are also susceptible, and any efforts to store gametes to preserve animal genetic resources would have to take cognizance of this fact (Easterday, 1965; Meegan et al., 1979).

Vesicular Stomatitis Virus

Vesicular stomatitis viruses (Indiana and New Jersey serotypes) are confined to the Americas. Although some understanding of the epidemiological behavior of the virus group is available, there are great gaps in essential knowledge. Virtually no studies have been carried out on transmission of the virus through the reproductive tract.

Domestic species of animals affected by these viruses include cattle, horses, and pigs; humans are also affected. Some wild animal species, including deer, are affected. The virus propagates readily in laboratory animals, such as rabbits, guinea pigs, and mice. A carrier state for these viruses has not been demonstrated and hence, the serological tests used by U.S. authorities to introduce zebu cattle from Brazil (House et al., 1983) were standardized to reject recently infected animals. Perhaps similar tests should be set for semen donors. This agent is one of the remaining enigmas of veterinary science and pathogenesis, and epidemiological studies are urgently needed because animals and gametes from the Americas constitute a large segment of international trade in livestock germplasm.

Contagious Bovine Pleuropneumonia

Contagious bovine pleuropneumonia (*Mycoplasma mycoides* subsp. *mycoides*) is a disease of cattle present in Africa, Portugal, Spain, occasionally other parts of Europe, Asia minor, some east European countries, and China.

Very little is known of the reproductive disorders caused by this microorganism. In studies carried out in Australia before the disease was eradicated in 1968, there was no indication of semen contamination, nor of intrauterine infection of the fetus (Seddon, 1965).

Swine Vesicular Disease Virus

Swine vesicular disease virus (SVDV) and (FMDV) have both been isolated from boar semen. Sows inseminated with SVDV seroconverted (McVicar and Eisner, 1977), thus demonstrating the potential to transmit SVDV infection through this route. Although difficulty with mounting was seen in boars affected with FMDV, it may not prevent collection of semen contaminated with FMDV or SVDV before manifestation of clinical disease (McVicar, 1984; Thacker et al., 1984).

African Swine Fever Virus

African swine fever virus (ASFV) has been described as causing a highly contagious, fatal disease in domestic pigs, but it also can be manifested as a mild to moderately severe disease, almost indistinguishable clinically from hog cholera. The virus is present in Africa, Italy, Portugal, Sardinia, and Spain. At present, ASFV appears to have been eradicated from parts of South America and the Caribbean. A report that ASFV, present in cryopreserved semen collected during viremia from an experimentally infected boar, was transmitted to a recipient female and produced infection (Schlafer, 1983) indicates the need to treat semen as a possible source of infection with this agent.

Hog Cholera Virus

Successful eradication programs in several countries have reduced the incidence of hog cholera virus (HCV) in herds of pigs. Because the most serious chronic manifestations of HCV are on reproductive conditions, including fertility, abortions, teratogenic defects, and neonatal deaths, control of this disease has widespread benefits (Cheville and Mengeling, 1969).

Evidence that a country is free of HCV is required for international movement of swine semen. Serological tests and disease control practices are becoming required conditions for members of the European Economic Community. Outbreaks of hog cholera are still occurring in Europe, however, as evidenced by an outbreak in the United Kingdom (Anonymous, 1986).

Summary

The diseases of greatest concern in the international movement of semen are those discussed above. Generally, the high pathogenicity and economic risks resulting from an infection by these agents ensure that production of gametes for international trade will be from countries free of these diseases. Increased risks may be involved and may be unavoidable, however, to procure access to highly desirable genetic material. Countries that largely depend on livestock industries have had to develop protocols to enable collection of semen and embryos from areas where diseases on the OIE's List A occur. Because of advances in testing methods for the agents listed and better disease controls within the countries where the agents are present, semen and gametes from high-risk areas of Africa and South America can now be transferred through offshore quarantine facilities to the United States, Australia, and New Zealand. Such transfers of gametes have increased prospects for international movement of animal germplasm and also have improved prospects for cryopreservation and use of rare, potentially desirable genetic materials.

List B Agents

The diseases on the OIE's List B are generally diseases for which adequate test systems are available in most countries. They are considered to be of socioeconomic or public health importance within countries and are significant in the international trade of livestock and livestock products. Some of these agents are described below.

Bacterial and Protozoal Agents

As stated by Wierzbowski (1984), the methods of testing bulls in artificial breeding centers for brucellosis, tuberculosis, paratuberculosis, leptospirosis, and trichomoniasis are well established and effective. Bartlett (1981) defined the need for uniformity of diagnostic methods for the diseases listed above as a basis for semen derived from specific pathogen-free donors, and Wierzbowski (1984) listed the principal diseases of concern for artificial breeding. With modern diagnostic tests, all the agents listed can be detected readily. For several of the diseases, either vaccines or reasonably adequate antibiotic treatment methods are available.

For campylobacteriosis, diagnostic tests that can identify carrier bulls in AB centers are available (Clark, 1985a). A vaccine developed in Australia can ensure eradication of campylobacteriosis from AB centers (Clark et al., 1974, 1976, 1979). Bovine trichomoniasis can be readily diagnosed (Clark, 1985b), and under some circumstances controlled or eradicated from AB centers (Clark et al., 1983). For both of the above venereal infections, diagnostic systems have been standardized in Australia. In addition, the vaccination of bovine donors and teasers in AB centers in Australia will help ensure freedom from *Campylobacter fetus* and its variants. There are no apparent reasons why these techniques cannot be used in other countries, with the same beneficial results of keeping donor bulls free of *C. fetus* and without the doubtful reliance on antibiotics in semen or in the treatment of carrier bulls.

Investigations on campylobacteriosis in donor cattle in Austria were reported by Flatscher and Holzman (1985) at the Fifty-Third General Session of the OIE. Perhaps as a result of these and similar studies, campylobacteriosis will cease to be a disease of concern in international trade and for long-term storage of germplasm.

Mycoplasmas

Examinations of bovine semen carried out in a series of surveys over recent years have noted the presence of a number of mycoplasma and ureaplasma contaminants. *Mycoplasma bovigenitalium* is usually the predominant organism along with ureaplasma. Other isolates, however, such as *M. bovis, M. canadense, M. californicum, M. alkalescens,* and *Mycoplasma* species group 7 have been recognized as pathogens in the bovine (Ball et al., 1987). Fresh semen (332 samples), processed semen (137 samples), and preputial washes yielded mycoplasmas or ureaplasmas from 46 percent, 31 percent, and 80 percent, respectively, and stored, processed semen straws taken as long ago as 1975 contained *M. canadense* and *M. bovigenitalium* (Ball et al., 1987). Addition of lincomycin and spectinomycin to the semen extender eliminated the capability to detect mycoplasmas and reduced the isolation of ureaplasmas.

The role of *M. bovis* infections of the bovine genital tract has not been clearly defined (Hartman et al., 1964; Hirth and Nielsen, 1966), although the infection is found in the reproductive tract at a lower incidence than *M. bovigenitalium, M. canadense,* or ureaplasmas (Friis and Blom, 1983). However, *M. bovigenitalium* has been found as a genital tract infection in bulls (Al-Aubaidi et al., 1972), and the disease has been reproduced experimentally (Parsonson et al., 1974). This has allowed identification of the acute and chronic inflammatory response to this organism, an agent often present among the microbiological contaminants of the preputial cavity. There have been several reports of the effect of mycoplasmas in artificial breeding and, in particular, their effect on fertility and freezability of bovine semen (Bartlett et al., 1976; Fejes, 1986; Kissi et al., 1985; Nielsen et al., 1987). In a report to the Committee on Infectious Diseases of Cattle of the U.S. Animal Health Association (Osburn, 1986), it was recommended that the antibiotic additives to control bacterial contamination in processed bull semen be replaced with gentamicin, lincomycin, spectinomycin and tylosin, specifically for mycoplasma. Certified Semen Services has adopted similar recommendations, and in order to comply with the CSS minimum requirements, the new antibiotic formula was to be phased in by January 1, 1988 (Doak, 1986). Although these requirements apply to artificial breeding in the United States, in order to supply germplasm to the U.S. market, countries around the world will have to adopt these standards, and so international trade will follow the dictates of the market (Doak, 1986).

The changes in antibiotic cover for bovine semen should be extended to buck and ram semen, for which the mycoplasma-induced diseases are of much greater importance. Jones (1983) reviewed the mycoplasmas of sheep and goats and identified those of proven pathogenicity. Because the urogenital tract is frequently colonized, it is important to minimize the effect of these organisms in artificial breeding. Mycoplasmas isolated from boar semen have been reviewed by Thacker et al. (1984), who concluded that the effect of these mycoplasma species in swine semen is poorly understood.

Because of the difficult growth requirements for bacterial isolation of mycoplasmas, isolation and identification from boar semen are difficult. However, the addition of the antibiotics lincomycin and spectinomycin would have an effect on controlling mycoplasmas in boar semen.

Infectious Bovine Rhinotracheitis

Because infectious bovine rhinotracheitis (IBR) viruses are so readily transmitted in semen and already found in cattle populations worldwide, importing countries usually require evidence of donor freedom from infection. Some strains of IBR, particularly those in North America, appear to be abortogenic, whereas strains in Australia and New Zealand normally do not cause abortions. Bovine herpesvirus 1 (BHV 1) causes mild to severe syndromes and often varies in clinical manifestations.

Because the BHV 1 group is virtually impossible to eradicate from the general cattle population, steps have been taken in Australia and Britain to maintain IBR-free bull herds in AB centers. Extensive testing of bovine semen to detect IBR virus, using a semen culture test (Sheffey and Krinsky, 1973) or a system devised by Kahrs and Littell (1980), has shown the difficulty of detecting minimal amounts of BHV 1.

Inoculation of semen contaminated with IBR virus into the uterus of cattle at estrus is followed by infection within days and seroconversion to the virus (Kendrick and McEntee, 1967; Parsonson and Snowdon, 1975). If semen batches are suspected of being contaminated with IBR virus and require testing, this may be the most reliable method at present. Loewen and Darcel (1985) compared two methods of isolating IBR virus from either milk or egg-yolk citrate semen extenders: centrifugation and dilution of samples with phosphate-buffered saline (pH 7.4). Both systems improved isolation of the virus.

Animal inoculation, either parentally or through the uterus at estrus, appears to be the most reliable method of detecting IBR (bovine herpesvirus 1) virus in semen (Parsonson and Snowdon, 1975; Schultz et al., 1982). The use of seronegative donors would reduce the likelihood of transmission and would be preferable to all other options.

Bovine Leukosis Virus

Testing cattle herds for bovine leukosis virus (BLV) antibodies has identified carrier herds in many countries. Using the reliable glycoprotein antigen in the agar gel immunodiffusion test, carrier bulls can be identified and not used as donors for international transfer of semen. There is a strong body of opinion, however, that leukocyte-free bovine semen will not transfer virus. Monke (1986) examined the use of bull semen originating from seven AB centers in the United States in a closed Jersey breed herd of 200 cows that had remained serologically negative for BL virus antibodies for 5 years. During this time, 24 of the 66 donor bulls were consistently serologically positive and a further 2, which provided 1,019 units (48.3 percent) of the semen used, became seropositive for BLV during the study.

Bovine Viral Diarrhea Virus

During evaluations of a procedure to test large numbers of semen samples for viral contamination, using an in vivo system to test pooled semen samples, called the Cornell semen test, Schultz et al. (1982) found that the only consistent viral contaminant present at infectious levels was bovine viral diarrhea virus (BVDV). The authors suggested that it would not be necessary for bulls to have overt clinical infections; subclinical infections are common.

Over a 4-year period, 40,000 samples from ejaculates were tested. No IBRV, bovine herpes mammillitis virus, BLV, or BTV were detected at levels sufficient to cause infection. The significance of the finding of BVDV in semen had not then been established (McClurkin et al., 1979; Whitmore et al., 1978); however, studies by Littlejohns (1982), McClurkin et al. (1984), and Roeder and Harkness (1986) have since established BVDV as a major viral disease of reproduction. Bulls, rams, or bucks persistently infected with BVDV can shed virus in semen. Semen contaminated with the virus could then introduce BVDV infection into closed cattle, sheep, or goat herds and cause subsequent losses in production through reproductive failure (Littlejohns, 1982; Lucas, 1986; Roeder and Harkness, 1986). The production of cattle immunotolerant to BVDV (McClurkin et al., 1984) highlights the ease with which this disease could spread in a noninfected herd. Control strategies for avoiding fetal infection in the early stages of gestation were included by Roeder and Harkness (1986) in a discussion of prospects for control of BVD viral infections. In a series of letters to the editor of the *Veterinary Record* (Brownlie et al., 1984; Harkness et al., 1984), control systems were discussed and evaluated. Control and eradication of the major viral diseases in bulls in artificial breeding centers, as in Britain, offer the most pragmatic solution (Lucas, 1986).

Ephemeral Fever Virus

Ephemeral fever of bovines and buffaloes is known to occur in many countries of southeast Asia, Africa, and Australia. In epizootics that have occurred in Australia, the morbidity in cows and heifers was approximately 90 percent, and in bulls 67 percent (Spradbrow and Francis, 1969), and temporary infertility was reported in bulls with clinical infection (Chenoweth and Burgess, 1972). In experimentally infected bulls, spermatozoa were found to have detached heads and tails, or bent and coiled tails, and ephemeral fever virus (EFV) was isolated from the semen of one bull. Ten cows inseminated at estrus with semen contaminated with EFV did not become infected, but remained susceptible to a subsequent challenge with EFV 30 days after insemination (Parsonson and Snowdon, 1974).

Akabane Virus

Parental inoculation of Akabane virus in eight bulls was followed by clinical, virological, and serological studies. Semen were collected regularly and examined for the presence of Akabane virus by inoculation of cell cultures and subcutaneous inoculation of susceptible cattle. Akabane virus was not detected in the semen by either isolation method. From the results of this study, it was concluded that Akabane viral infection of the bull would not affect reproduction (Parsonson et al., 1981a).

Parainfluenza 3 Virus

An ubiquitous virus in cattle and sheep populations, parainfluenza 3 (PI-3) virus, has been isolated from bull semen on occasion. Abraham and Alexander (1986) reported isolation of PI-3 in a group of 15 bulls one month after vaccination with a live PI-3 vaccine, but they could not detect any relationship between this virus and the fertility of the bulls.

Porcine Parvovirus

The reproductive losses associated with parvovirus infections in susceptible pregnant sows and gilts (young females) have been shown to cause major problems in piggeries (Joo and Johnson, 1976; Mengeling, 1979). Moreover, the infection is usually attributed to viral contamination of the environment, either by direct contact through the oral-nasal route or through the recently introduced pigs or from contaminated instruments. Bonte et al. (1984) experimentally introduced porcine parvovirus through the oral-nasal route into boars and found evidence of the virus in the genital tract before immunity was induced. They concluded that it would be possible to introduce infection through semen.

To avoid transmission of infectious agents through boar semen, constraints and disease testing of boars in special centers developed for the purpose of artificial breeding would be required, much as they are in bovine AB centers. Control measures ensuring freedom from specific porcine pathogens have been called for in pig-breeding centers in the United States and in the United Kingdom (Thacker et al., 1984). The same high standards of freedom from specific diseases and maintenance of such freedom by a disease-testing program, quarantine and testing of new introductions, and strict attention to hygiene in the swine AB centers would be essential. By using the same approach with swine AB centers as that adopted for controlling bovine semen, similarly high-quality standards for semen could be attained. Trying to assess the microorganism contamination of semen is very unrewarding because of the difficulty of culturing semen from animals. Inoculation into animals, as in the Cornell semen test, offers a possible means for monitoring contamination, but it has a very low benefit-cost relationship in comparison with achieving disease freedom from important pathogens through vaccination and the hygienic handling, preservation, and storage of boar semen.

Diseases of Sheep and Goats

The major disease of sheep and goats that is currently inhibiting the international movement of semen from these species is scrapie. The countries of the world that are free of scrapie do not want to introduce the disease. Australia, New Zealand, and South Africa are among the only major sheep-producing countries of the world that are free of both scrapie and maedi-visna.

Recently, efforts have been made in Australia and New Zealand to introduce new sheep and goat genotypes using systems based on artificial breeding and long-term quarantine (up to 5 years). Early results of experimental trials in the United States with scrapie-infected breeding stock and noninfected recipients have been reported by Foote et al. (1986). The experiments consist of groups in which embryo transfer in sheep with artificially induced scrapie embryo transfer in goats with artificially induced scrapie and embryo transfer in sheep with naturally occurring scrapie are being conducted. After 5 years of the trial, only the positive control group (both donors and recipients were scrapie exposed), have expressed the disease.

An additional project is the artificial insemination of scrapie-free ewes with semen from scrapie-infected sires. After 3 years, there has been no evidence of scrapie in the progeny. If these results continue, artificial breeding may provide the means to control this disease and to ensure the safe collection of genetic material from sheep and goats in countries where scrapie is a problem.

For the long-term storage of germplasm from sheep and goats, such assurances as freedom from scrapie, maedi-visna, and caprine arthritis and encephalitis would be mandatory. Scrapie has become the most difficult of all animal diseases to diagnose and control because of the inability to diagnose it except on clinical signs, which then have to be followed by histological examination and reproduction of the disease in homologous and heterologous species based on intracerebral inoculation of brain and other tissues.

In those countries in which scrapie does not exist, the opportunities for a structured AB program are present. Development of such programs has taken place in Australia and New Zealand, often at existing cattle AB centers. The standard disease controls are similar to those established for cattle. Requirements for quality control, handling, storage, and transport of semen are also similar to those for cattle. Now, with the movement to storage of ram and buck semen in ministraws instead of pellets, there is less problem with contamination in storage. Because none of the diseases of sheep and goats caused by the retroviruses of maedi-visna and pulmonary adenomatosis in sheep or by caprine arthritis and encephalitis in goats, has been shown to be transmitted by artificial breeding, use has been made of AB to control these diseases in countries where they are present.

Diseases of Horses

Although there are some severe technical problems with artificial breeding of horses, there are requirements for superior genetic material, notwithstanding the restrictions on AB by some equine breed societies. Without the inhibition of breed-governing groups that oppose AB, the techniques for artificial breeding and preservation of gametes will evolve. Most AB in horses is carried out within standard-bred horse societies and by some companion and pleasure horse groups, but AB is not generally used for commercial purposes and is still not widely accepted.

A major bacterial venereal disease of horses, contagious equine metritis (CEM), caused by *Haemophilus equigenitalis* (Taylor et al., 1978), appeared in Thoroughbred horses in several countries of Europe (Timoney et al., 1977), in North America (Swerczek, 1978), and in Australia and New Zealand in 1977 (Hughes et al., 1978). The spread of the agent was traced to the movement of several stallions used for breeding, and it highlights how rapidly such agents can be dispersed with the aid of modern transportation and commercialization. Because of the value of the Thoroughbred industry, many millions of dollars can be lost when mares are infertile for a breeding season. The stallion with CEM is an asymptomatic carrier and can transmit infection for extended periods of time. Although CEM has been eradicated from the horse populations in Australia, New Zealand, North America, Britain, and Ireland, there is a continuing need for quarantine vigilance against reintroduction of this disease agent.

A number of viral diseases of horses are arthropod borne, for example, African horse sickness, equine encephalosis, Venezuelan equine encephalitis and Eastern and Western equine encephalitis, vesicular stomatitis, and equine infectious anaemia. There is little or no evidence for the involvement of the equine encephalitides in equine reproduction, other than the disease state of the animals during breeding.

Several equine viral diseases have a marked effect on reproduction, including the equine herpes viruses, EHV 1, EHV 2, and EHV 3.

Of these, EHV 1 is a major pathogen of horses; it causes respiratory disease (subtype 2) and abortion (subtype 1).

Equine arteritis has been shown to infect some stallions persistently, and the route of spread can be by venereal transmission from semen (Timoney et al., 1986).

There are several viral diseases of horses for which there is insufficient information to identify clearly the role the causal agents play in reproduction. The agents include the equine influenzas, A equine 1 (H7N7) and A equine 2 (H3N8), Borna disease virus, Japanese encephalitis virus, and Getah virus.

An increasing demand for cryopreserved equine semen has led to more effort being devoted to developing better semen extender systems (Francl et al., 1987) and to improving the methods of freezing semen to ensure preservation of fertility in equine semen (Volkmann, 1987).

Poultry Diseases

The intensification and integration of the poultry industry has concentrated large flocks of birds, both broilers and layers, within relatively small areas. These concentrations of birds require a very high level of disease control. Most preventive methods are based on quarantine of important birds, close disease monitoring, and the development of specific pathogen-free breeder units, especially elite and grandparent flocks. Even in such flocks, great care must be taken in regard to the use of preventive vaccines, food and water supplies, and general hygiene, as well as isolation from environmental factors (Allan, 1981).

The major pathogens of concern are those that are known to be transmitted by the vertical route. Diseases of concern include *Salmonella pullorum, Mycoplasma gallisepticum, M. synoviae,* and *M. meleagridis,* and the viral diseases of infectious laryngotracheitis, adenovirus 127, and avian leukosis complex. In the majority of elite and grandparent flocks, these diseases have been eradicated and are regularly monitored. Eradication of all diseases from the commercial areas is too difficult, and in these flocks, vaccines play a large role in disease control (Allan, 1981).

For artificial breeding of poultry, disease control in the donor flocks is essential. For long-term preservation of germplasm, it will be imperative to use only specific pathogen free flocks. For preservation of endangered or rare bird species, a high standard of disease control will be necessary, but because the genetic material will be rare, some accommodation may have to be made.

The cryopreservation of chicken semen from inbred and specialized strains will be a vital resource for specific pathogen free elite and grandparent flocks. Storage of semen from these sources will provide a safeguard in the event of a change in disease status. More important, however, it will provide a valuable genetic library for the future. Studies by Sexton (1979), Lake et al. (1981), Lake (1986), and Bacon et al. (1986) have established the possibilities for long-term cryopreservation of chicken semen. Established methods for artificial breeding of poultry will enable storage of inbred and elite lines of chickens and other birds, including endangered wild species.

ARTIFICIAL BREEDING OF ENDANGERED AND RARE SPECIES

The numbers of rare and endangered species of animals and birds are rapidly increasing as humans move into the breeding ranges of these species. The importance of utilizing all knowledge available from artificial breeding to attempt the continuation of endangered species and, where possible, the long-term preservation of their gametes for posterity, is a vital issue for this generation (Durrant et al., 1986; Mason, 1974).

The development of animal model systems for embryo technologies for rare and endangered species of wildlife has generated great interest among several concerned groups (Wildt et al., 1986). The Convention on International Trade in Endangered Species of Wild Flora and Fauna (U.S. Department of the Interior, 1984) has recognized more than 200 mammalian species as being threatened by extinction. The accelerated rate of habitat destruction is leaving conservationists very little choice other than managing most wildlife in captive situations, such as in zoological gardens or parks and wildlife reservations. Many of the wildlife species that are threatened require very careful, expert handling and treatment to ensure that the techniques adopted for their preservation are suitable and not harmful or stressful to their well-being.

Because of the lack of knowledge of the genetic background of many endangered species, with respect to the size of the genetic base, any information that is pertinent is urgently required. This also applies to reproductive technologies, including gamete collection, handling, storage, and treatment. Until the information is available and organized into developing strategies for propagation of species by artificial means, there is little chance of using AB as a viable alternative. Yet, it offers the only major hope for the preservation of many species.

RECOMMENDATIONS FOR FURTHER RESEARCH

The OIE has set standards for the hygienic collection and handling of fresh and preserved bovine semen in artificial breeding centers, appendix 5.2.1.1 (1986). The OIE states that observation of the standards will result in bovine semen almost free of common bacteria.

In the OIE code for bovine semen (1986) from AB centers accredited for export, the purpose of official sanitary control in semen production is to maintain the health of animals at an AB center at a standard that permits the international distribution of semen free of specific pathogenic organisms that can be carried in semen and cause infection in recipient female cattle. In discussing the cryopreservation of genetic material and the disease risks and controls, Wierzbowski (1984) pointed out the danger of contaminated single ejaculates being spread over a long time period. Hare (1985) emphasized the advantages of semen as a disease-control method over animal-to-animal contact. Collection of semen can be made from donors known to be free of specific pathogens, and a statistically defined sample of any individual collection can be tested for freedom from specific microorganisms before insemination. An additional benefit is that the donors can be tested for evidence of previous infection for periods of time, extending into months if necessary, before their semen is used in artificial breeding.

Hare (1985) elaborates on the historical development of AB centers and the development of legislative controls based on country of origin, herds, and finally individual donors and their semen. Such controls have become recognized internationally.

Proposals for freedom from specific pathogenic microorganisms for bovine semen were made originally by Bartlett et al. (1976). However, some countries, such as Australia, already had a strict code for the production of semen free of specific pathogens. Disease-free requirements are now being recommended internationally, and the disease lists generally adopted are based on the OIE's A, B, and Other lists. It is expected that in a member country, donor animals producing semen will be free of diseases on List A, the AB herd should also be free of diseases on List B, and the individual donors should be free of diseases on the Other list (Office International des Épizooties, 1986).

Although the OIE has been cited as having an influence on AB, Wierzbowski (1984) has correctly suggested that the national or state veterinary authorities should be responsible for the health status of

AB centers producing germplasm for international trade. In addition, the Food and Agriculture Organization of the United Nations should produce lists of suitable AB centers, with annual updates to establish standards of quality internationally. Irrespective of the official body that undertakes the task, continuous monitoring of the diseases to be excluded and notification of newly recognized pathogenic microorganisms would be essential to the effectiveness of such standards. The international trade in germplasm has functioned with exceptional efficiency, based on the requirements of the importing countries and the standards for AB centers of the exporting countries. Only very rarely is transmission of disease in germplasm recognized as occurring following international transfer (Kupferschmied et al., 1986). When such cases do occur in a country in which the disease has been controlled or eradicated, the consequences for the disease control authorities of the recipient country can be extremely difficult, and the country may incur considerable financial costs in reinstituting freedom from the disease.

Diagnostic tests for many diseases likely to be transmitted by semen are being improved, and new tests are being developed in many laboratories engaged in this research around the world. The problems of searching for agents in semen are manifold, however, and have been the subject of extensive research. Such methods are far too costly and often are grossly inefficient when compared with testing donor animals before and after collection of semen and prior to release of semen (Kahrs and Littell, 1980; Parez, 1984; Richmond et al., 1977; Schultz et al., 1982). For the majority of viral diseases in domestic animals, there is a concurrent viremia whenever the virus is excreted in semen, and therefore, it is preferable to test for antibodies or virus in the blood. The recent development of technologies, such as the polymerase chain reaction (Erlich, 1989), will enable screening of semen and embryo samples for evidence of specific pathogens and should enhance the confidence in international movement of germplasm.

For microorganisms present in the preputial cavity of animals that are commonly transferred into the semen collection, other than viral or protozoal contaminants, the dilution of semen in extenders tends to lower the probability of infection. In addition, the use of antibiotics provides a control on the proliferation of bacterial and fungal contaminants. For the diseases for which vaccination can be used as a prophylactic measure, such as campylobacteriosis, excretion of immunoglobulins into the preputial cavity enables control of the infection. Similarly, if vaccines are used to protect against or control general infections, secreted immunoglobulins will reduce the possibility of semen contamination. In some tropical environments, the climate and the normal behavior of animals can make it difficult to control microorganisms, for example, water buffalo and bucks in India (Gangadhar et al., 1986; Sharma and Deka, 1986).

In AB centers, there will always be populations of microorganisms, many of which will reside in the preputial sheath and on the surface of the penis. The aims of the OIE recommendations and the majority of AB centers are to confine and restrict pathogenic microorganisms. Many of the major bacterial diseases of concern in the first AB centers, including brucellosis, tuberculosis, leptospirosis, campylobacteriosis, and trichomoniasis, have now been eradicated or are controlled, and are no longer considered problem diseases. New disease agents have emerged, however, such as mycoplasmas, ureaplasmas, and *Haemophilus* spp. In addition, some of the viral diseases not considered previously are now recognized as potentially spread through semen. The most important viral disease in this category is BVDV, and some attempts are now being made to eliminate bulls carrying this agent from AB centers in Britain (Lucas, 1986).

Pestiviruses also cause a disease of great importance in sheep flocks (Border disease), and thus, ram semen should be free of this group of viruses. Bulls that have been infected in utero with BVDV and appear clinically normal and test seronegative will frequently excrete the virus (Littlejohns, 1982). Although BVDV has been found in semen (McClurkin et al., 1979), it is very difficult to culture from semen and could be present as a contaminant. Thus, BVD seronegative bulls in AB centers could best be checked by attempting to isolate the virus from blood and by carrying out serological testing to ensure freedom from BVDV.

Infectious bovine rhinotracheitis virus is known to be a contaminant of semen when there is a recrudescence of the disease on the penis and prepuce of infected bulls (Snowdon, 1964; Studdert et al., 1964). Once bulls have exhibited penile infection, the disease can intermittently recur with shedding of virus. Bulls that are seronegative for IBR virus are the only animals which can reliably be used for the production of semen for export (Kupferschmied et al., 1986).

Following intranasal IBR virus vaccination of bulls and testing large numbers of semen samples in the Cornell semen test (Schultz et al., 1982), IBRV was not detected in the semen of IBR seropositive bulls. The authors suggested that there may be a variation in the excretion sites between vaccinated (intranasal) and naturally infected bulls.

Because of the detection of bluetongue virus in the semen of bulls (Bowen et al., 1983; Breckon et al., 1980; Luedke et al., 1975; Parsonson et al., 1981b), and the possibility of a bull remaining persistently infected and excreting virus intermittently in his semen (Luedke et al., 1975), many countries have required a seronegative status for BTV from all donor bulls. At present, this requirement is essential to ensure semen free of BTV. There are indications, however, that this requirement may in fact be far too rigid and restrict access to desirable germplasm because experimental and field experience has not verified the "carrier state" for BTV (Monke et al., 1986; Parsonson et al., 1987b).

Bovine leukosis (BL) virus has not been found to be shed in semen (Schultz et al., 1982). Monke (1986) found that seropositive bulls infected with BLV did not transmit the virus in their semen.

Ideally, bulls at AB centers should be specific pathogen free, particularly when storing semen for long periods of time is being considered (Bartlett, 1981). However, because not all viral diseases produce a "carrier state" in bulls, serologic evidence to show that there has not been a rise in antibody titer, but rather that there has been a fall in antibody titer, may be sufficient to enable collection of semen from seropositive bulls. With BVD virus, seropositive bulls, or seronegative bulls from whom virus could not be isolated from the blood, could be used for the long-term storage of semen. For IBR virus, seronegative animals are preferable, but, as Schultz et al. (1982) showed, under some circumstances even these animals could be used. Obviously, a specific virus free status for the important viral agents, as is required in many AB centers in Britain (Lucas, 1986) and Australia, is the ideal status to achieve.

Within the United States, the Certified Semen Services provides a voluntary code, *Minimum Requirements for Health of Bulls Producing Semen for A.I.*, for the member organizations of the National Association of Animal Breeders to establish standards for herd health and standards of hygiene for AB establishments (Howard, 1986). Although these standards are proposed for bull AB centers, the principles would be applicable for other species of domestic animals used in artificial breeding.

REFERENCES

Abraham, A., and R. Alexander. 1986. Isolation of parainfluenza-3 virus from bull's semen. Vet. Rec. 119:502.

Al-Aubaidi, J. M., K. McEntee, D. H. Lein, and S. J. Roberts. 1972. Bovine seminal vesiculitis and epididymitis caused by *Mycoplasma bovigenitalium*. Cornell Vet. 62:581–596.

Allan, W. H. 1981. Virus diseases of poultry. Pp. 205–216 in Virus Diseases of Food Animals, Vol. 1, E. P. J. Gibbs, ed. London: Academic Press.

Anonymous. 1986. Classical swine fever in Shropshire. Vet. Rec. 118:438.

Australian Quarantine and Inspection Service. 1988. Minimum Health Standards for Stock Standing at Licensed or Approved Artificial Breeding Centers in Australia, 3d ed. Canberra: Australian Government Publishing Service.

Bacon, L. D., D. W. Salter, J. V. Motta, L. B. Crittenden, and F. S. Ogasawara. 1986. Cryopreservation of chicken semen of inbred or specialized strains. Poult. Sci. 65:1965–1971.

Ball, H. J., E. F. Logan, and W. Orr. 1987. Isolation of mycoplasmas from bovine semen in Northern Ireland. Vet. Rec. 121:322–324.

Bartlett, D. E., L. L. Larson, W. G. Parker, and T. H. Howard. 1976. Specific pathogen free (SPF) frozen bovine semen: A goal? Proc. Tech. Conf. A. I. Reprod. 6:11–22.

Bartlett, D. E. 1981. Bull semen: Specific micro-organisms. Pp. 29–48 in Disease Control in Semen and Embryos. FAO Animal Production and Health Paper 23. Rome, Italy: Food and Agriculture Organization of the United Nations.

Bonte, P., M. Vandeplassche, and P. Biront. 1984. Porcine parvovirus in boars. II. Influence on fertility. Zentralbl. Veterinaermed. B. 31:391–395.

Bowen, R. A., and T. H. Howard. 1984. Transmission of bluetongue virus by intrauterine inoculation or insemination of virus-containing bovine semen. Am. J. Vet. Res. 45:1386–1388.

Bowen, R. A., T. H. Howard, K. W. Entwistle, and B. W. Pickett. 1983. Seminal shedding of bluetongue virus in experimentally infected mature bulls. Am. J. Vet. Res. 44:2268–2270.

Breckon, R. K., A. J. Luedke, and T. E. Walton. 1980. Bluetongue virus in bovine semen: Viral isolation. Am. J. Vet. Res. 41:439–442.

Brownlie, J., M. C. Clarke, and C. J. Howard. 1984. Mucosal disease in cattle. Vet. Rec. 115:158.

Cheville, N. F., and W. L. Mengeling. 1969. The pathogenesis of chronic hog cholera (swine fever). Histologic, immunofluorescent and electron microscopic studies. Lab. Invest. 20:261–274.

Chenoweth, P. J., and G. W. Burgess. 1972. Mid-piece abnormalities in bovine semen following ephemeral fever. Aust. Vet. J. 48:37–38.

Clark, B. L. 1985a. Bovine *campylobacteriosis*. Pp. 1–5 in Australian Standard Diagnostic Techniques for Animal Diseases, No. 22. Melbourne, Australia: Commonwealth Scientific and Industrial Research Organization.

Clark, B. L. 1985b. Bovine trichomoniasis. Pp. 1–4 in Australian Standard Diagnostic Techniques for Animal Diseases, No. 23. Melbourne, Australia: Commonwealth Scientific and Industrial Research Organization.

Clark, B. L., J. H. Dufty, M. J. Monsborough, and I. M. Parsonson. 1974. Immunization against bovine vibriosis. 4. Vaccination of bulls against infection with *Campylobacter fetus* subsp. *venerealis*. Aust. Vet. J. 50:407–409.

Clark, B. L., J. H. Dufty, M. J. Monsborough, and I. M. Parsonson. 1976. Immunization against bovine vibriosis due to *Campylobacter fetus* subsp. *fetus* biotype *intermedius*. Aust. Vet. J. 52:362–365.

Clark, B. L., J. H. Dufty, M. J. Monsborough, and I. M. Parsonson. 1979. A dual vaccine for immunization of bulls against vibriosis. Aust. Vet. J. 55:43.

Clark, B. L., J. H. Dufty, and I. M. Parsonson. 1983. Immunization of bulls against trichomoniasis. Aust. Vet. J. 60:178–179.

Cottral, G. E., P. Gailuinas, and B. F. Cox. 1968. Foot-and-mouth disease virus in semen of bulls and its transmission by artificial insemination. Arch. Gesamte Virusforsch. 23:362–377.

Doak, G. A. 1986. Certified Semen Services implementation of new antibiotic combination. Pp. 42–43 in Proceedings of the 11th Technical Conference on Artificial Insemination and Reproduction, Milwaukee, Wisconsin, April 25–26, 1986.

Durrant, B. S., J. E. Oosterhuis, and M. L. Hoge. 1986. The application of artificial reproduction techniques to the propagation of selected endangered species. Theriogenology 25:25–32.

Easterday, B. C. 1965. Rift Valley fever. Adv. Vet. Sci. 10:65–127.

Erlich, H. A., ed. 1989. PCR Technology Principles and Applications for DNA Amplification. New York: Stockton.

Fejes, J. 1986. Pathological changes in the genital system of breeding bulls with mycoplasmas in semen. Veterinarstvi 36:118–119.

Fenner, F. 1976. Classification and nomenclature of viruses. Intervirology 6:1–12.

Flatscher, J., and A. Holzman. 1985. Genital disease in bulls: Importance for artificial insemination-control measures. Pp. 393–402 in Proceedings of the 11th Conference of the Office des International Épizooties, Regional Commission for Europe, Vienna, September 25–28, 1984.

Foote, W. C., J. W. Call, T. C. Bunch, and J. R. Pitcher. 1986. Embryo transfer in the control of transmission of scrapie in sheep and goats. Proc. Annu. Meet. U. S. Anim. Health Assoc. 90:413–416.

Francl, A. T., R. P. Amann, E. L. Squires, and B. W. Pickett. 1987. Motility and fertility of equine spermatozoa in a milk extender after 23 or 24 hours at 20°C. Theriogenology 27:517–526.

Friis, N. F., and E. Blom. 1983. Isolation of *Mycoplasma canadense* from bull semen. Acta Vet. Scand. 24:315–317.

Gangadhar, K. S., A. R. Rao, and S. Krishnaswamy. 1986. Bacterial load in the frozen semen of buffalo bulls. Indian Vet. J. 63:320–324.

Gibbs, E. P. J., ed. 1981. Virus Diseases of Food Animals, Vols. 1 and 2. New York: Academic Press.

Hare, W. C. D. 1985. Diseases Transmissible by Semen and Embryo Transfer Techniques. Technical Series No. 4. Paris: Office International des Épizooties.

Harkness, J. W., L. Wood, and T. Drew. 1984. Mucosal disease in cattle. Vet. Rec. 155:283.

Hartman, H. A., M. E. Tourtellotte, S. W. Nielsen, and W. N. Plastridge. 1964. Experimental bovine uterine mycoplasmosis. Res. Vet. Sci. 5:303–310.

Hirth, R. S., and S. W. Nielsen. 1966. Experimental pathology of bovine salpingitis due to mycoplasma insemination. Lab. Invest. 15:1132–1133.

House, J. A., Yedloutschnig, R. J., A. H. Dardiri, D. E. Herrick, and J. A. Acree. 1983. Procedures used for the importation of Brazilian zebu cattle into the United States. Proc. Am. Assoc. Vet. Lab. Diag. 26:13–24.

Howard, T. H. 1986. CSS sire health: Present and future. Proc. Tech. Conf. A. I. Reprod. 11:19–26.

Hughes, K. L., J. D. Bryden, and F. McDonald. 1978. Equine contagious metritis. Aust. Vet. J. 54:101.

Jones, G. E. 1983. Mycoplasmas of sheep and goats: A synopsis. Vet. Rec. 113:619–620.

Joo, H. S., and R. H. Johnson. 1976. Porcine parvovirus: A review. Vet. Bull. 46:653–660.

Kahrs, R. F., and R. C. Littell. 1980. Detection of viruses in bovine semen. Proc. Am. Assoc. Vet. Lab. Diag. 23:251–261.

Kendrick, J. W., and K. McEntee. 1967. The effect of artificial insemination with semen contaminated with IBE-IPV virus. Cornell Vet. 57:3–11.

Kissi, B., S. Juhasz, and L. Stipkovits. 1985. Effect of mycoplasma contamination of bull semen on fertilization. Acta Vet. Hung. 33:107–117.

Kitching, R. P., and W. P. Taylor. 1985. Clinical and antigenic relationship between isolates of sheep and goat pox viruses. Trop. Anim. Health Prod. 17:64–74.

Kitching, R. P., J. M. Hammond, and W. P. Taylor. 1987. A single vaccine for the control of capripox infection in sheep and goats. Res. Vet. Sci. 40:53–60.

Kupferschmied, H. U., U. Kihm, P. Bachmann, K. H. Muller, and M. Ackermann. 1986. Transmission of IBR/IPV virus in bovine semen: A case report. Theriogenology 25:439–443.

Lake, P. E. 1986. The history and future of the cryopreservation of ovine germ plasm. Poult. Sci. 65:1–15.

Lake, P. E., O. Ravie, and J. McAdam. 1981. Preservation of fowl semen in liquid nitrogen: Application to breeding programmes. Br. Poult. Sci. 22:71–77.

Littlejohns, I. R. 1982. Bovine mucosal disease and its familial virus. Pp. 665–667 in Viral Diseases in South-East Asia and the Western Pacific, J. S. McKenzie, ed. Sydney, Australia: Academic Press.

Loewen, K. G., and Darcel, C. le Q. 1985. A comparison of two methods for the isolation of bovine herpesvirus 1 (BHV-1) from extended bovine semen. Theriogenology 23:935–943.

Lucas, M. H. 1986. Control of virus diseases in bulls in artificial insemination centres in Britain. Vet. Rec. 119:15–16.

Luedke, A. J., T. E. Walton, and R. H. Jones. 1975. Detection of bluetongue virus in bovine semen. Proc. 20th World Vet. Cong. 38:2039–2042.

Luedke, A. J., M. M. Jochim, and R. H. Jones. 1977. Bluetongue in cattle: Effects of *Culicoides variipennis* transmitted bluetongue virus on pregnant heifers and their calves. Am. J. Vet. Res. 38:1687–1695.

MacLachlan, N. J., C. E. Schore, and B. I. Osburn. 1984. Antiviral responses of bluetongue virus-inoculated bovine fetuses and their dams. Am. J. Vet. Res. 45:1469–1473.

Mason, I. L. 1974. Conservation of animal genetic resources. Pp. 13–94 in First World Congress on Genetics Applied to Livestock Production, Vol. 2. Madrid, Spain: Graficas Orbe.

McClurkin, A. W., M. F. Coria, and R. C. Cutlip. 1979. Reproductive performance of apparently healthy cattle persistently infected with bovine viral diarrhea virus. J. Am. Vet. Med. Assoc. 174:1116–1119.

McClurkin, A. W., E. T. Littledike, R. C. Cutlip, G. H. Frank, M. F. Coria, and S. R. Bolin. 1984. Production of cattle immunotolerant to bovine viral diarrhea virus. Can. J. Comp. Med. 48:156–161.

McVicar, J. W. 1984. Boar semen. Pp. 70–79 in Proceedings International Symposium on Microbiological Tests for the International Exchange of Animal Genetic Material, O. H. V. Stalheim, ed. Columbia, Mo.: American Association of Veterinary Laboratory Diagnosticians.

McVicar, J. W., and R. J. Eisner. 1977. Foot-and-mouth disease and swine vesicular disease viruses in boar semen. Proc. Annu. Meet. U. S. Anim. Health Assoc. 81:221–230.

Meegan, J. M., H. Hoogstraal, and M. I. Moussa. 1979. An epizootic of Rift Valley fever in Egypt in 1977. Vet. Rec. 105:124–125.

Mengeling, W. L. 1979. Prenatal infection following maternal exposure to porcine parvovirus on either the seventh of fourteenth day of gestation. Can. J. Comp. Med. 43:106–109.

Monke, D. R. 1986. Noninfectivity of semen from bulls infected with bovine leukosis virus. J. Am. Vet. Med. Assoc. 188:823–826.

Monke, D. R., W. D. Hueston, and J. W. Call. 1986. Veterinary management of an artificial insemination center containing bluetongue seropositive bulls. Proc. Annu. Meet. U. S. Anim. Health Assoc. 90:139–143.

Nielsen, K. H., R. B. Stewart, M. M. Garcia, and M. D. Eaglesome. 1987. Enzyme immunoassay for detection of *Mycoplasma bovis* antigens in bull semen and preputial washings. Vet. Rec. 120:596–598.

Odend'hal, S., ed. 1983. The Geographical Distribution of Animal Viral Diseases. New York: Academic Press.

Office International des Épizooties. 1986. International Zoo-Sanitary Code, 5th ed. Paris: Office International des Épizooties.

Osburn, B. I. 1986. Report of the Committee on Bluetongue and Bovine Leukosis. Proc. Annu. Meet. U. S. Anim. Health Assoc. 90:154–159.

Parez, M. 1984. The most important genital diseases of cattle control, treatment and hygiene of semen collection. Pp. 175–203 in Proceedings of the 11th Conference of the Office International des Épizooties, Regional Commission for Europe, Vienna, September 25–28, 1984. Paris: Office International des Épizooties.

Parsonson, I. M. 1990. Pathology and pathogenesis of bluetongue infections. Pp. 119–141 in Current Topics in Microbiology and Immunology, Vol. 162, P. Ross and B. M. Gorman, eds. Berlin: Springer-Verlag.

Parsonson, I. M., and W. A. Snowdon. 1974. Ephemeral fever virus: Excretion in the semen of infected bulls and attempts to infect female cattle by the intrauterine inoculation of virus. Aust. Vet. J. 50:329–334.

Parsonson, I. M., and W. A. Snowdon. 1975. The effect of natural and artificial breeding using bulls infected with, or semen contaminated with, infectious bovine rhinotracheitis virus. Aust. Vet. J. 51:365–369.

Parsonson, I. M., J. M. Al-Aubaidi, and K. McEntee. 1974. *Mycoplasma bovigenitalium:* Experimental induction of genital disease in bulls. Cornell Vet. 64:240–264.

Parsonson, I. M., A. J. Della-Porta, W. A. Snowdon, and M. L. O'Halloran. 1981a. Experimental infection of bulls with Akabane virus. Res. Vet. Sci. 31:157–160.

Parsonson, I. M., A. J. Della-Porta, D. A. McPhee, D. H. Cybinski, K. R. E. Squire, H. A. Standfast, and M. F. Uren. 1981b. Isolation of bluetongue virus serotype 20 from the semen of an experimentally infected bull. Aust. Vet. J. 57:252–253.

Parsonson, I. M., A. J. Della-Porta, D. A. McPhee, D. H. Cybinski, K. R. E. Squire, and M. F. Uren. 1987a. Bluetongue virus serotype 20: Experimental infection of pregnant heifers. Aust. Vet. J. 64:14–17.

Parsonson, I. M., A. J. Della-Porta, D. A. McPhee, D. H. Cybinski, K. R. E. Squire, and M. F. Uren. 1987b. Experimental infection of bulls and cows with bluetongue virus serotype 20. Aust. Vet. J. 64:10–13.

Plowright, W. 1968. Rinderpest virus. Pp. 27–110 in Virology Monographs 3, S. Gard, C. Hallauer, and K. F. Meyer, eds. New York: Springer-Verlag.

Richmond, J. V., J. W. McVicar, R. J. Eisner, L. A. J., Johnson, and V. G. Pursel. 1977. Diagnostic problems for virus detection in boar semen. Proc. Annu. Meet. U. S. Anim. Health Assoc. 81:231–243.

Roeder, P. L., and J. W. Harkness. 1986. BVD virus infection: Prospects for control. Vet. Rec. 118:143–147.

Schlafer, D. H. 1983. Boar semen. Pp. 65–69 in Proceedings of an International Symposium on Microbiological Tests for the International Exchange of Animal Genetic Material, O. H. V. Stalheim, ed. Columbia, Mo.: American Association of Veterinary Laboratory Diagnosticians.

Schultz, R. D., L. S. Adams, G. Letchworth, B. E. Sheffey, T. Manning and B. Bean. 1982. A method to test large numbers of bovine semen samples for viral contamination and results of a study using this method. Theriogenology 17:115–123.

Scott, G. R. 1964. Rinderpest. Pp. 113–223 in Advances in Veterinary Science, Vol. 9, C. Brandly, and E. L. Jungherr, eds. New York: Academic Press.

Seddon, H. R. 1965. Bacterial diseases. Pp. 1–142 in Diseases of Domestic Animals in Australia, Vol. 2, Pt. 5, H. E. Albiston, ed. Canberra, Australia: Commonwealth of Australia, Department of Health.

Seidel, G. E., Jr. 1986. Impact of biotechnology on the future of artificial insemination. Proc. Tech. Conf. A. I. Reprod. 11:96–101.

Sellers, R. F., R. Burrows, J. A. Mann, and P. Dawe. 1968. Recovery of virus from bulls affected with foot-and-mouth disease. Vet. Rec. 83:303.

Sexton, T. J. 1979. Studies on the fertility of frozen fowl semen. Eighth Int. Cong. Anim. Reprod. Artific. Insem. Krakow 4:1079–1082.

Sharma, D. K., and B. C. Deka. 1986. Bacterial flora in frozen buck semen. Indian J. Comp. Microbiol. Immunol. Infect. Dis. 7:40–42.

Sheffey, B. E., and M. Krinsky. 1973. Infectious bovine rhinotracheitis virus in extended bovine semen. Proc. Annu. Meet. U. S. Anim. Health Assoc. 77:131–137.

Shin, S. 1986. The control of mycoplasmas, ureaplasmas, *Campylobacter fetus* and *Haemophilus somnus* in frozen bovine semen. Proc. Tech. Conf. A. I. Reprod. 11:33–38.

Snowdon, W. A. 1964. Infectious bovine rhinotracheitis and infectious pustular vulvovaginitis in Australian cattle. Aust. Vet. J. 40:277–288.

Spradbrow, P. B., and J. Francis. 1969. Observations on bovine ephemeral fever and isolation of virus. Aust. Vet. J. 45:525–527.

Studdert, M. J., C. A. V. Barker, and M. Savan. 1964. Infectious pustular vulvovaginitis virus infection of bulls. Am. J. Vet. Res. 25:303–314.

Swerczek, T. W. 1978. Contagious equine metritis in the U. S. A. Vet. Rec. 102:512–513.

Taylor, C. E. D., R. O. Rosenthal, D. F. J. Brown, S. P. Lapage, L. R. Hill, and R. M. Legros. 1978. The causative organism of contagious equine metritis 1977: Proposal for a new species to be known as *Haemophilus equigenitalis*. Equine Vet. J. 10:136–144.

Thacker, B. J., R. E. Larson, H. S. Joo, and A. D. Leman. 1984. Swine diseases transmissible with artificial insemination. J. Am. Vet. Med. Assoc. 185:511–516.

Timoney, P. J., J. Ward, and P. Kelly. 1977. A contagious genital infection of mares. Vet. Rec. 101:103.

Timoney, P. J., W. H. McCollum, A. W. Roberts, and T. W. Murphy. 1986. Demonstration of the carrier state in naturally acquired equine arteritis virus infection in the stallion. Res. Vet. Sci. 41:279–280.

U.S. Department of the Interior. 1984. The Convention on International Trade in Endangered Species of Wild Flora and Fauna. Washington, D.C.: U.S. Fish and Wildlife Service.

Volkmann, D. H. 1987. Acrosomal damage and progressive motility of stallion semen frozen by two methods in 0.5 milliliter straws. Theriogenology 27:689–698.

Weiss, K. E. 1968. Lumpy skin disease virus. Pp. 111–131 in Virology Monographs, Vol. 3, S. Gard, C. Hallauer, and K. F. Meyer, eds. New York: Springer-Verlag.

Whitmore, H. L., B. K. Gustafsson, P. Havareshti, A. B. Duchateau, and E. C. Mather. 1978. Inoculation of bulls with bovine virus diarrhea virus: Excretion of virus in semen and effects on semen quality. Theriogenology 9:153–163.

Wierzbowski, S. 1984. Cryopreservation of genetic material: Disease risk and control. Pp. 49–65 in Animal Genetic Resources: Cryogenic Storage of Germplasm and Molecular Engineering. FAO Animal Production and Health Paper No. 4412. Rome, Italy: Food and Agriculture Organization of the United Nations.

Wildt, D. E., M. C. Schiewe, P. M. Schmidt, K. L. Goodrowe, J. G. Howard, L. G. Phillips, S. J. O'Brien, and M. Bush. 1986. Developing animal model systems for embryo technologies in rare and endangered wildlife. Theriogenology 25:33–51.

Glossary

allele One of two or more alternative forms of a gene, differing in DNA sequence and affecting the functioning of a single gene product (RNA and/or protein). Any of the alternate allele forms may occupy the same site or locus on a pair of homologous chromosomes.

allelic frequency The percentage of all alleles at a given locus in a population gene pool represented by a particular allele.

artificial selection A form of selection that is superimposed on natural selection by breeders in an effort to change various traits in preferred stocks.

backcross A cross of heterozygous offspring with individual genotypically identical to one of its parents.

biodiversity (biological diversity) The variety and variability among living organisms and the ecological complexes in which they occur.

blastocyst The mammalian embryo at the time of its implantation into the uterine wall.

blastomere A cell produced during the splitting of a fertilized egg.

breed For the purposes of this report and in general, an interbreeding group with some identifiable common appearance, performance, ancestry, selection history, or other feature.

chromosome A thread-like structure of the cell's nucleus that carries genetic information arranged in a linear sequence.

clone A group of genetically identical cells or organisms all descended from a single common ancestral cell or organism.

conservation The management of human use of the biosphere so that it may yield the greatest sustainable benefit to present generations while maintaining its potential to meet the needs and aspirations of future generations.

crossbreeding The mating members of two genetically distinct populations. In the context of this report, mating among two or more different breeds.

cryobiology The study of the effects of extremely low temperature on biological systems.

cryopreservation Maintaining tissues, semen, eggs, or embryos for the purpose of long-term storage at ultralow temperatures, typically between –150°C and –196°C; produced by storage above or in liquid nitrogen.

cytogenetics The combined study of cells and genes at the chromosome level.

data base Refers to a system—usually computer-based—in which information is compiled in an organized, retrievable manner.

diploid Possessing twice the number of chromosomes as the number present in reproductive cells such as eggs. The somatic number of chromosomes (2N).

DNA Deoxyribonucleic acid. Each DNA molecule has two strands of complementary nitrogenous base pairs structurally arranged in a double helix. It is the molecular basis of heredity in many organisms.

DNA hybridization A technique for determining the degree of similarity of DNA base sequences in two species, involving radioisotope labeling of DNA from one species and annealing it to the DNA of the other species.

DNA sequencing A method for determining the sequence of base pairs in sections of DNA.

ecosystem A community of organisms interacting with one another; the environment in which organisms live and with which they also interact.

effective population size The estimated number of individuals in a population that contribute genes to succeeding generations, accounting for factors such as unequal sex ratio of breeding individuals, variation in number of offspring per parent, and changes in breeding population size over time.

electrophoresis The differential movement of charged molecules in solution through a porous medium in an electric field. The porous medium can be filter paper, cellulose, or, more frequently, a starch or polyacrylamide gel.

electrophoretic analysis A method commonly used to separate proteins and other organic molecules.

embryo The early or developing stage of an organism, especially the developing product of egg fertilization.

embryo transfer Artificial introduction of a recently formed embryo into the oviduct or uterus of the biological mother or of a surrogate mother.

endangered In the context of this report, a term that applies to taxa (species, subspecies, population) in danger of extinction and for which survival is unlikely if the causal factors of loss continue.

enzyme One of a large group of proteins, produced by living cells, that act like catalysts in essential chemical reactions in living tissues.

exotic Relating to an organism that is found in, or is intentionally introduced into, an area where it does not naturally occur.

extensive management The production and management system under which domestic livestock has traditionally been raised, generally under range conditions. Nutritional supplementation and veterinary care are minimal or nonexistent. Animals are subjected to a variety of environmental and nutritional stresses and become adapted to their local environments.

extinct In the context of this report refers to taxa (e.g., populations, subspecies, species) not found after repeated searches of known and likely areas.

fecundity Potential fertility or the capability of repeated fertilization. Specifically, it is the quantity of eggs, produced per female, over a defined period of time.

feral Pertaining to formerly domesticated animals now living in a wild state.

flock Temporary management units, or groupings of animals, on a given farm that are not necessarily maintained in reproductive isolation and may not be genetically different.

flow cytometry A method for rapidly measuring cells.

gamete A mature reproductive cell (egg or sperm) that is capable of fusing with another similar cell of the opposite sex to produce a zygote (fertilized egg).

gene The basic functional unit of inheritance that occupies a specific position within a chromosome or genome and is responsible for the heritability of particular traits.

gene flow The movement of genes through or between populations as the result of crossbreeding and natural selection.

gene mapping Assignment of a gene locus to a specific chromosome within a genome and/or determining the sequence of genes and their relative distances from one another on a specific chromosome.

gene pool The totality of genes and their alleles within an interbreeding population.

gene transfer Refers to the identification and isolation of a gene from one genome and its transfer into the genome of another individual. Transfers may be within the same species or from one species to another.

genetic conservation To maintain a reservoir of potentially useful genetic variation from that contained in the multitude of breeds and varieties.

genetic distance A measure of the number of allelic substitutions per locus that have occurred during the separate evolution of two populations or species.

genetic diversity In a population or species, the possession of a variety of genetic traits that frequently result in differing expressions in different individuals.

genetic drift The variation of allele frequencies from one generation to the next due to random fluctuations. In small populations this can lead to significant changes.

genetic resources Organisms from which the genes needed by breeders and other scientists can be derived.

genetic risk In the context of this report, the potential for an activity or event to result in loss of genes from a species or population.

genome A single complete set of the genes or chromosomes of an individual. Typically, gametes such as egg cells contain a single set and are termed haploid, while the somatic cells that comprise the bulk of the living tissue of the plant body contain two sets and are diploid.

genomic library A collection of transgenic DNA molecules (cloned fragments) that is representative of individual genomes.

genotype The genetic constitution of an organism, as distinguished from its appearance. The genotype may or may not be expressed, in appearance or performance, depending on the environmental effects of a given location. *See* phenotype.

genotype-environment interaction Where the phenotypic expression of a genotype differs when measured under different environmental conditions.

germplasm The tissue, semen, eggs, embryos, or juvenile or mature animals useful in breeding, research, and conservation efforts.

germplasm bank An institution or center that participates in the management of genetic resources, in particular by maintaining ex situ or in situ collections; the term also can refer to a collection of genetic resources rather than the institution holding it.

germplasm collection A collection of many different varieties, species, or subspecies that represent a diverse collection of genotypes and, hence, genetic diversity.

habitat The place where an organism is usually found; its natural environment.

herd *See* flock.

herdbooks Records of pedigrees and breeding stocks selected for performance that began to appear in eighteenth-century Europe.

heterosis Increased vigor in terms of growth, fertility, or other characters in individuals which resulted from a cross of two genetically different populations.

heterozygous Having one or more unlike alleles at corresponding loci of homologous chromosomes.

homologous chromosomes Chromosomes in a cell that have the same visible structure and linear sequences of genes.

homozygous Having like alleles for a particular gene at corresponding loci on homologous chromosomes.

hybrid The product of a cross between two species, races, or genetically distinct individuals.

hybridization The process of crossing individuals that possess different genetic makeups.

in vitro fertilization Fertilization of an egg outside the female body.

inbreeding The mating of related individuals.

industrialized stocks Livestock reared in uniform, high-quality conditions so that minimal selection effort is necessary for environmental adaptation. Often breeders begin to develop one highly selected strain for specific production traits, and it comes to dominate production.

intensive management Production and management systems for domestic livestock that involve high levels of inputs, such as supplemental feed and veterinary care and require greater management effort.

isozyme (isoenzyme) Different chemical forms of the same enzyme that can generally be distinguished from one another by electrophoresis.

kilobase (kb) A measure of nucleic acid length consisting of 1,000 nucleotides.

landrace A population typically genetically heterogeneous, that is specifically adapted to a set of local conditions.

line A term used in the livestock breeding industry to indicate a group of individuals of a given breed that are maintained in reproductive isolation from other members of the breed to develop potentially advantageous production traits.

linkage group A group of genes that are in close proximity on the same chromosome and, therefore, are generally inherited together.

livestock Domesticated animals that are generally associated with various types of agricultural production and are used as sources of draft power, food, or fiber. They include cattle, horses, pigs, chickens, sheep, and goats.

locus (plural, **loci**) The position that a gene occupies in a chromosome.

major domesticates The 40 domesticated mammalian and avian species that are used extensively on a global scale due to their adaptability and utility and because cultural preferences for them have developed.

minor domesticates The domesticated mammalian and avian species that are less abundant globally than major domesticates but are critically important because of their relative abilities to produce in environmental niches and because of traditional or cultural patterns of use for their products.

mitochondrial DNA In animal cells, DNA found outside of the nucleus in small, oblate bodies called mitochondria.

morula An embryo that consists of a cluster of cleaving blastomeres; a stage prior to the blastula.

multiple ovulation An animal reproduction technology involving the hormonal treatment of the donor female to induce ovulation and to produce greater numbers of embryos per donor female.

mutation A gene that is modified as a result of structural change. The individual that results may be referred to as a mutant. The term also can be applied to the process producing such genetic change.

natural selection The differential fecundity in nature between members of a species possessing adaptive characters and those without such advantages.

oocyte An egg before fertilization.

outcross the mating of genetically unrelated individuals from genetically different accessions, breeding lines, populations, or species.

ovum A mature egg; female gamete. This term can also refer to a preimplantation embryo.

phenotype The sum total of the environmental and genetic (hereditary) influences on an organism; the visible or measurable characteristics of an organism. The degree to which these characteristics become manifest is often determined by the interaction of the alleles that comprise the genotype with environmental factors. *See* genotype.

polymorphism Having more than one form or structural pattern. In genetics this can refer to the existence within a population of several alternative alleles at a single locus.

population A group of organisms of the same species that occupy a particular geographic area or region. In general, individuals within a population potentially interbreed with one another.

population genetics The study of the genetic composition and dynamics of populations.

preservation That aspect of conservation by which a sample of an animal genetic resource population is designated to an isolated process of maintenance, by providing an environment free of the material and human forces that might bring about genetic change.

progeny testing The evaluation of the genotype of a parent by a study of its progeny under controlled conditions.

quantitative genetics The branch of genetics that is concerned with measurable traits determined by several pairs of genes. *See* quantitative traits.

quantitative traits Traits determined by the cumulative action of many genes, usually in concert with environmental factors.

rare In the context of this report, taxa with small populations that are not currently endangered but that are at risk of loss.

reproductive isolation The inability of two populations or species to produce fertile offspring or interbreed as a consequence of physiological or genetic incompatibility or environmental barriers.

restriction endonuclease Any one of many enzymes that cleave DNA molecules at specific recognition sites. They are coded for by genes called restriction alleles.

restriction fragment length polymorphism Variation that occurs within a species in the length of DNA fragments resulting from digestion of the extracted DNA with one of several enzymes that cleave DNA at specific recognition sites. Changes in the genetic composition result in fragments of altered length.

selection Any natural or artificial process that permits an increase in the proportion of certain genotypes or groups of genotypes in succeeding generations in relation to others.

semen immunoextension A method of adding antibodies to specific antigens of the pathogens potentially found in semen to control specific diseases and prevent transmission in the semen.

semen screening A method for verifying the health status of semen samples.

species A taxonomic subdivision; a group of organisms that actually or potentially interbreeds and is reproductively isolated from other groups.

stable A system that returns to its initial conditions after being disturbed.

stock *See* line.

strain *See* line.

tandem repeat Two identical chromosomal segments that lie one behind the other. The order of the genes in each segment is the same.

transgenic animal An animal into which cloned genetic material has been experimentally transferred.

type A distinction defined by the production characteristics of livestock species. For example, there are two types of cattle, meat and dairy; two types of sheep, hair and wool; and two types of chickens, layers and broilers. Many different breeds can be found within a given type.

unique Having distinct characteristics that are not found in another breed or type.

wild species Organisms in or out of captivity that have not been subject to breeding to alter them from their native (wild) state.

Abbreviations

AB	Artificial breeding
ALPA	Asociación Latinoamericana de Producción Animal (Latin American Association of Animal Production)
AMBC	American Minor Breeds Conservancy, United States
CAB	Commonwealth Agricultural Bureaux, England
CENARGEN	Centro Nacional de Recursos Genéticos (National Center for Genetic Resources), Brazil
CGIAR	Consultative Group on International Agricultural Research, United States
COAG	Committee on Agriculture (FAO)
DIMDI	Deutsches Institut für Medizinische Dokumentation und Information (German Institute for Medical Documentation and Information), Germany
DNA	deoxyribonucleic acid
EAAP	European Association for Animal Production, Italy
EMBRAPA	Empresa Brasileira de Pesquisa Agropecuária (Brazilian Agricultural Research Corporation), Brazil
FAO	Food and Agriculture Organization of the United Nations, Italy
IBAR	Inter-African Bureau for Animal Resources, Kenya
IBPGR	International Board for Plant Genetic Resources, Italy
ILCA	International Livestock Center for Africa, Ethiopia

ILRAD	International Laboratory for Research on Animal Diseases, Kenya
IUCN	International Union for Conservation of Nature and Natural Resources, Switzerland
MHC	major histocompatibility complex
mtDNA	mitochondrial DNA
N_e	effective population size
OAU	Organization of African Unity, Ethiopia
^{32}**P**	a radioactive isotope of phosphorus
RBST	Rare Breeds Survival Trust, England
RFLP	restriction fragment length polymorphism
SABRAO	Society for the Advancement of Breeding Researches in Asia and Oceania, Malaysia
UNEP	United Nations Environment Program, Kenya
VNTR	variable number of tandem repeats

Authors

ROBERT W. ALLARD Allard is emeritus professor of genetics at the University of California, Davis. He has a Ph.D. degree in genetics from the University of Wisconsin. His areas of research include plant population genetics, gene resource conservation, and plant breeding. He is a member of the National Academy of Sciences.

PAULO DE T. ALVIM Since 1963 Alvim has been the scientific director for the Comissão Executiva do Plano da Lavoura Cacaueira, Brazil. He earned a Ph.D. degree from Cornell University with specialization in plant physiology, tropical agriculture, and ecology.

JOHN H. BARTON Barton joined the Stanford law faculty in 1969, and is now George E. Osborne Professor of Law and director of the school's International Center on Law and Technology. He is a Stanford law graduate, and teaches and consults extensively on international technology law issues, particularly those associated with agricultural biotechnology, intellectual property rights, and genetic resources.

FREDERICK H. BUTTEL is professor of rural sociology at the University of Wisconsin, where he earned a Ph.D. degree in sociology. He is also adjunct professor of science and technology studies at Cornell University. His areas of interest are in technology and social change, particularly in relation to agricultural research and biotechnology.

TE-TZU CHANG Chang is formerly head of the International Rice Germplasm Center (1983–1991) and principal geneticist (1985–1991) at the International Rice Research Institute, Philippines. He has been a visiting professor at the University of the Philippines, Los Baños, since 1962. He has a Ph.D. degree in plant genetics and breeding from the University of Minnesota. Chang had a vital role in the Green Revolution in rice and has broad experience in managing and designing plant gene banks. He is president of the Society for the Advancement of Breeding Researches in Asia and Oceania.

PETER R. DAY (*Committee Chair*) Before joining Rutgers University as director of the Center for Agricultural Molecular Biology in 1987, Day was the director of the Plant Breeding Institute, Cambridge, United Kingdom. He has a Ph.D. degree from the University of London, and is a leader in the field of biotechnology and its application to agriculture.

ROBERT E. EVENSON Since 1977 Evenson has been a professor of economics at Yale University. He has a Ph.D. degree in economics from the University of Chicago. His research interests include agricultural development policy with a special interest in the economics of agricultural research.

HENRY A. FITZHUGH (*Subcommittee Chair*) Fitzhugh is deputy director general for research at the International Livestock Center for Africa, Ethiopia. He received a Ph.D. degree in animal breeding from Texas A&M University. His field of research is the development and testing of biological and socioeconomic interventions to improve the productivity of livestock in agricultural production systems.

MAJOR M. GOODMAN Goodman is professor of crop science, statistics, genetics, and botany at North Carolina State University (NCSU) where he has been employed since 1967. He has a Ph.D. degree in genetics from NCSU, and his areas of research are plant breeding, germplasm conservation and utilization, numerical taxonomy, history and evolution of maize, and applied multivariate statistics. Goodman is a member of the National Academy of Sciences.

JAAP J. HARDON In 1985 Hardon became the director of the Center for Genetics Resources, The Netherlands. He has a Ph.D. degree in plant genetics from the University of California. His specialty is plant breeding and genetics.

ELIZABETH L. HENSON Since 1986 Henson has been a director of the Cotswold Farm Park's Rare Breeds Survival Center in the United Kingdom. Prior to this she served as executive director for the American Minor Breeds Conservancy and conducted its first census of breeds in North America. She has a master's degree in zoology from the University of Oxford and a master's degree in animal breeding from the University of Edinburgh.

JOHN HODGES Until August 1990 Hodges was senior officer for animal breeding and genetic resources with the Food and Agriculture Organization of the United Nations. He has a Ph.D. degree in animal genetics from the University of Reading and is a graduate of the Harvard Business School. Formerly professor of animal genetics at the University of British Columbia, Vancouver, Canada, he now lives in Austria and is working as a private consultant and writer in the fields of animal production, breeding and genetics, biotechnology, biodiversity, and environmental biology.

DONALD R. MARSHALL In 1991 Marshall left the Waite Agricultural Research Institute at the University of Adelaide, Australia, to accept a position as professor of plant breeding at the University of Sydney. He has a Ph.D. degree in genetics from the University of California, Davis. His professional interests are population genetics, plant breeding, host-parasite interactions, and genetic resources conservation.

DAVID R. NOTTER Since 1977 Notter has been on the faculty of Virginia Polytechnic Institute and State University, where he is currently professor of animal science. He has M.S. and Ph.D. degrees in animal breeding and genetics from the University of Nebraska. His research has been in the development of breeding and management systems in beef cattle and sheep, and the genetic control of animal reproduction with emphasis on the breeding season of sheep.

DIETER PLASSE Plasse is professor of genetics in the Faculty of Veterinary Science at the Universidad Central de Venezuela. He has a Dr. Sc. Agr. degree (Doctor in Agriculture Science) from the Georg-August-Universität of Göttingen, Germany. His research involves studies of breeding and selection to improve productivity of beef cattle populations in the Latin American tropics using *Bos indicus* (Criollo) and *Bos taurus* (including Criollo) cattle.

SETIJATI SASTRAPRADJA Sastrapradja is affiliated with the National Center for Research in Biotechnology at the Indonesian Institute of Science. She has a Ph.D. degree in botany from the University of Hawaii.

LOUISE LETHOLA SETSHWAELO Setshwaelo is the senior agricultural research officer for animal breeding in the Ministry of Agriculture, Botswana. She has a Ph.D. degree in animal breeding and genetics from the University of Nebraska. Her work is in crossbreeding, selection, and breed development of cattle in southern Africa.

CHARLES SMITH Smith is a professor of animal breeding strategies at the University of Guelph, Canada. He has a Ph.D. degree in animal breeding from Iowa State University. His research area is in animal breeding strategies, including genetic conservation, and he has been involved in international efforts to conserve domestic animal germplasm.

JOHN A. SPENCE In 1989 Spence was appointed head of the Cocoa Research Unit at the University of the West Indies, Trinidad and Tobago. He has a Ph.D. degree from the University of Bristol, United Kingdom. His research interests are cocoa tissue culture and cryopreservation as alternatives to holding field germplasm collections.

THOMAS E. WAGNER Since 1984 Wagner has been director of the Edison Animal Biotechnology Center at Ohio University. He has a Ph.D. degree in biochemistry from Northwestern University. His current research interests include control mechanisms in the regulation of genetic expression, genetic recombination in eucaryotes, mammalian recombinant gene transfer, and biotechnology.

JAMES E. WOMACK Womack is the W. P. Luse Endowed Professor of the Department of Veterinary Pathology in the College of Veterinary Medicine, Texas A&M University. He has a Ph.D. degree in genetics from Oregon State University. His research interests include biochemical and comparative genetics of mammals, the physiological significance of isozyme and DNA polymorphisms, animal models of human genetic disease, developmental genetics, and organization of the mammalian genome.

Index

C

D

E

F

G

S

Recent Publications of the Board on Agriculture

Policy and Resources

Sustainable Agriculture and the Environment in the Humid Tropics (1993), 720 pp., ISBN 0-309-04749-8.

Agriculture and the Undergraduate: Proceedings (1992), 296 pp., ISBN 0-309-04682-3.

Water Transfers in the West: Efficiency, Equity, and the Environment (1992), 320 pp., ISBN 0-309-04528-2.

Managing Global Genetic Resources: Forest Trees (1991), 244 pp., ISBN 0-309-04034-5.

Managing Global Genetic Resources: The U.S. National Plant Germplasm System (1991), 198 pp., ISBN 0-309-04390-5.

Sustainable Agriculture Research and Education in the Field: A Proceedings (1991), 448 pp., ISBN 0-309-04578-9.

Toward Sustainability: A Plan for Collaborative Research on Agriculture and Natural Resource Management (1991), 164 pp., ISBN 0-309-04540-1.

Investing in Research: A Proposal to Strengthen the Agricultural, Food, and Environmental System (1989), 156 pp., ISBN 0-309-04127-9.

Alternative Agriculture (1989), 464 pp., ISBN 0-309-03985-1.

Understanding Agriculture: New Directions for Education (1988), 80 pp., ISBN 0-309-03936-3.

Designing Foods: Animal Product Options in the Marketplace (1988), 394 pp., ISBN 0-309-03798-0; ISBN 0-309-03795-6 (pbk).

Agricultural Biotechnology: Strategies for National Competitiveness (1987), 224 pp., ISBN 0-309-03745-X.

Regulating Pesticides in Food: The Delaney Paradox (1987), 288 pp., ISBN 0-309-03746-8.

Pesticide Resistance: Strategies and Tactics for Management (1986), 480 pp., ISBN 0-309-03627-5.

Pesticides and Groundwater Quality: Issues and Problems in Four States (1986), 136 pp., ISBN 0-309-03676-3.

Soil Conservation: Assessing the National Resources Inventory, Volume 1 (1986), 134 pp., ISBN 0-309-03649-9; Volume 2 (1986), 314 pp., ISBN 0-309-03675-5.

New Directions for Biosciences Research in Agriculture: High-Reward Opportunities (1985), 122 pp., ISBN 0-309-03542-2.

Genetic Engineering of Plants: Agricultural Research Opportunities and Policy Concerns (1984), 96 pp., ISBN 0-309-03434-5.

Nutrient Requirements of Domestic Animals Series and Related Titles

Nutrient Requirements of Horses, Fifth Revised Edition (1989), 128 pp., ISBN 0-309-03989-4; diskette included.

Nutrient Requirements of Dairy Cattle, Sixth Revised Edition, Update 1989 (1989), 168 pp., ISBN 0-309-03826-X; diskette included.

Nutrient Requirements of Swine, Ninth Revised Edition (1988), 96 pp., ISBN 0-309-03779-4.

Vitamin Tolerance of Animals (1987), 105 pp., ISBN 0-309-03728-X.

Predicting Feed Intake of Food-Producing Animals (1986), 95 pp., ISBN 0-309-03695-X.

Nutrient Requirements of Cats, Revised Edition (1986), 87 pp., ISBN 0-309-03682-8.

Nutrient Requirements of Dogs, Revised Edition (1985), 79 pp., ISBN 0-309-03496-5.

Nutrient Requirements of Sheep, Sixth Revised Edition (1985), 106 pp., ISBN 0-309-03596-1.

Nutrient Requirements of Beef Cattle, Sixth Revised Edition (1984), 90 pp., ISBN 0-309-03447-7.

Nutrient Requirements of Poultry, Eighth Revised Edition (1984), 71 pp., ISBN 0-309-03486-8.

Further information, additional titles (prior to 1984), and prices are available from the National Academy Press, 2101 Constitution Avenue, NW, Washington, DC 20418, 202/334-3313 (information only); 800/624-6242 (orders only); 202/334-2451 (fax).